Learning Mathematics for a New Century

LEARNING MATHEMATICS FOR A NEW CENTURY

Maurice J. Burke
2000 Yearbook Editor
Montana State University

Frances R. Curcio
General Yearbook Editor
New York University

2000 YEARBOOK

National Council of Teachers of Mathematics
Reston, Virginia

Copyright © 2000 by
THE NATIONAL COUNCIL OF TEACHERS OF MATHEMATICS, INC.
1906 Association Drive, Reston, VA 20191-9988
www.nctm.org
All rights reserved

Library of Congress Cataloging-in-Publication Data:

Learning mathematics for a new century / Maurice J. Burke, Frances R. Curcio, [editors].
    p. cm. — (Yearbook ; 2000)
  Includes bibliographical references and index.
  ISBN 0-87353-479-4
    1. Mathematics— Study and teaching. I. Burke, Maurice Joseph. II. Curcio, Frances R.
III. Yearbook (National Council of Teachers of Mathematics) ; 2000.

QA1 .N3 2000
[QA11]
510 s—dc21
[510′.71′273]

                                                                                                        00-027485

> The publications of the National Council of Teachers of Mathematics present a variety of viewpoints. The views expressed or implied in this publication, unless otherwise noted, should not be interpreted as official positions of the Council.

Printed in the United States of America

Contents

Preface ix

Prologue

1. Perspectives on Mathematics Education 1
   STEPHEN S. WILLOUGHBY
   University of Arizona, Tucson, Arizona

Part 1: Numeracy and Standards

2. The Four Faces of Mathematics 16
   KEITH DEVLIN
   Saint Mary's College of California, Moraga, California

3. The Many Roads to Numeracy 28
   DOROTHY WALLACE
   Dartmouth College, Hanover, New Hampshire

4. The Standards Movement in Mathematics Education: Reflections
   and Hopes 37
   JOAN FERRINI-MUNDY
   Michigan State University, East Lansing, Michigan

Part 2: Technology and the Mathematics Classroom

5. Calculators in Mathematics Teaching and Learning: Past, Present,
   and Future 51
   BERT K. WAITS
   Ohio State University, Columbus, Ohio
   FRANKLIN DEMANA
   Ohio State University, Columbus, Ohio

6. Technology-Enriched Learning of Mathematics: Opportunities
   and Challenges 67
   FRANK WATTENBERG
   Montana State University, Bozeman, Montana
   LEE L. ZIA
   University of New Hampshire, Durham, New Hampshire

7. Using Extranets in Fostering International Communities of
   Mathematical Inquiry 82
   LYN D. ENGLISH
   Queensland University of Technology, Brisbane, Queensland, Australia
   DONALD H. CUDMORE
   Oxford University, Oxford, England

Part 3: The School Mathematics Curriculum

8. As the Century Unfolds: A Perspective on Secondary School
   Mathematics Content 96
   JOHNNY W. LOTT
   University of Montana, Missoula, Montana
   TERRY A. SOUHRADA
   University of Montana, Missoula, Montana

9. The Impact of *Standards*-Based Instructional Materials in
   Mathematics in the Classroom 112
   ERIC E. ROBINSON
   Ithaca College, Ithaca, New York
   MARGARET F. ROBINSON
   Ithaca College, Ithaca, New York
   JOHN C. MACELI
   Ithaca College, Ithaca, New York

10. Beyond Eighth Grade: Functional Mathematics for
    Life and Work.. 127
    SUSAN L. FORMAN
    Bronx Community College, Bronx, New York
    LYNN ARTHUR STEEN
    Saint Olaf College, Northfield, Minnesota

11. Statistics for a New Century................................. 158
    RICHARD L. SCHEAFFER
    University of Florida, Gainesville, Florida

Part 4: Improving Mathematical Learning Environments

12. Supporting Students' Ways of Reasoning about Data 174
    KAY MCCLAIN
    Vanderbilt University, Nashville, Tennessee
    PAUL COBB
    Vanderbilt University, Nashville, Tennessee
    KOENO GRAVEMEIJER
    Freudenthal Institute, Netherlands, and
    Vanderbilt University, Nashville, Tennessee

13. Talking about Math Talk.................................... 188
    MIRIAM GAMORAN SHERIN
      Northwestern University, Evanston, Illinois
    EDITH PRENTICE MENDEZ
      Sonoma State University, Rohnert Park, California
    DAVID A. LOUIS
      The Nueva School, Hillsborough, California

14. Blending the Best of the Twentieth Century to Achieve a
    Mathematics Equity Pedagogy in the Twenty-first Century 197
    KAREN C. FUSON
      Northwestern University, Evanston, Illinois
    YOLANDA DE LA CRUZ
      Arizona State University—West, Phoenix, Arizona
    STEPHEN T. SMITH
      Northwestern University, Evanston, Illinois
    ANA MARÍA LO CICERO
      Northwestern University, Evanston, Illinois
    KRISTIN HUDSON
      Northwestern University, Evanston, Illinois
    PILAR RON
      Northwestern University, Evanston, Illinois
    REBECCA STEEBY
      Northwestern University, Evanston, Illinois

15. Exploring Mathematics through Talking and Writing............ 213
    DAVID J. WHITIN
      Queens College of the City University of New York, Flushing, New York
    PHYLLIS WHITIN
      Queens College of the City University of New York, Flushing, New York

Epilogue

16. Unfinished Business: Challenges for Mathematics Educators in the
    Next Decades 223
    JEREMY KILPATRICK
      University of Georgia, Athens, Georgia
    EDWARD A. SILVER
      University of Pittsburgh, Pittsburgh, Pennsylvania

Index 237

Preface

There I was, staring into my crystal ball. I saw that starting with a few choice axioms and many undefined terms and applying mathematical reasoning to the problems of the twentieth century, mathematics education had become a branch of applied mathematics by the middle of the twenty-first century. For example, take the controversy raging over calculators versus pencil and paper. Many educators in the twentieth century had claimed that knowing how to divide two- and three-digit numbers using a calculator was better than nothing. Others asserted with equal conviction that nothing was better than knowing how to do such divisions with paper and pencil. When the two axioms were put together using the transitive property of "better than," it was thereby proved that knowing how to divide on a calculator was better than knowing how to divide with paper and pencil.

With equal ease was proved a conjecture long assumed by twentieth-century mathematics educators but often ignored in the classroom. Named the fundamental theorem of mathematics education, the conjecture stated that "it can't be remembered," referring presumably to mathematics taught and learned strictly by rote. Unfortunately, the proof of this theorem had also been forgotten by the middle of the twenty-first century, with only the record of a tantalizing note scribbled in the margin of an old copy of the NCTM 2000 Yearbook testifying that, though easy, the proof was too large to fit in the margin of the book.

The image in the crystal ball faded as I came across the incompleteness theorem of mathematics education concerning something about the product-process controversy. All turned dark....

The 2000 Yearbook of the National Council of Teachers of Mathematics (NCTM) is titled *Learning Mathematics for a New Century* out of the desire to reflect on where we have been and to share our visions of what *might* be and what factors *might* shape that future. The call for articles did not ask authors to predict what *will* be. Few individuals would have been willing to write such articles unless science fiction, like the vision above, had been allowed. What the future will be is for the future to determine. However, many factors will contribute to the story of mathematics education in the twenty-first century. Many are discussed in the articles of this volume. One way of looking at this to-be-written story is captured in the conclusion to the article by Wattenberg and Zia: "The twenty-first century will be nothing if not exciting. Hop on board!"

This volume has a prologue, four major sections, and an epilogue. The prologue, by Stephen Willoughby, reflects nearly half a century of professional involvement in mathematics education, including a term as president of the NCTM. Willoughby introduces the major themes of the yearbook in his reflections on the various movements he has experienced within mathematics education. He brings up the themes of technology, equity, policy, and standards that reverberate throughout the volume. And, at the heart of it all, he argues that mathematics education in the twenty-first century must find a way to make learning mathematics instructive and rewarding for the student.

Part 1 of the yearbook contains three articles focusing on numeracy and standards. In a century when policy decisions will likely be made on such connected issues as school choice, school accountability, and school standards, we must reexamine our overall goals. In his article, "The Four Faces of Mathematics," Keith Devlin points out that numeracy must not be equated with only *mathematics education*. He asserts that numeracy, like literacy, should be the responsibility of all teachers in all lessons. The study of mathematics in school must be broadened to build appreciation for its true nature. The "skills" goals of our system must be brought in line with other equally important goals.

Dorothy Wallace's article, "The Many Roads to Numeracy," carries the Devlin article a step further in questioning what standards we should expect for everyone in his or her mathematical learning. Wallace's "ecological" model offers a unique perspective on a socially responsible balance in our mathematics education goals. Finally, Joan Ferrini-Mundy shares her perspective on the development of national standards in mathematics education. Providing an inside view of the updating of the original NCTM *Standards* documents, Ferrini-Mundy captures the tension in the details with which a nation must struggle in coming to terms with what mathematics education can or should be in the twenty-first century.

Part 2 of the yearbook focuses on technology and the mathematics classroom. After witnessing the explosive effects of technological change on nearly every aspect of twentieth-century society, few can doubt that technology will be a stimulus for change in mathematics classrooms of the twenty-first century. The Bert Waits and Franklin Demana article discusses the influence of calculators in making sophisticated mathematical software readily available to students in the twenty-first century. However, availability alone will not determine the outcome. An important mediating factor in the use of classroom technology is the preparation of the teacher. Waits and Demana and nearly every other author in the 2000 Yearbook discuss the crucial importance of teacher preparation. There is also a growing debate on what is "appropriate use" of computer and calculator technology in the learning of school mathematics. The debate often hinges on what we mean by basic skills and indeed on what skills should be basic in mathematics education. Waits and Demana offer their own perspective on the issue but recognize that the argument is far from over. (In the epilogue, Jeremy Kilpatrick and

Edward Silver pose this technology issue as one of the significant challenges for mathematics educators in the next decades.)

Frank Wattenberg and Lee Zia broaden the discussion of technology in the twenty-first century classroom to include the World Wide Web, scientific devices, and many other tools. Using the study of waves as a rich context for mathematical learning, they illustrate some of the plethora of tools available today and invoke those tools as pointers to the "tool space" of tomorrow. One of the principal challenges of the twenty-first century will be the management of this burgeoning tool space so that teachers and students can use it effectively in the learning of mathematics. They conclude their article with a vision of how a new kind of "library" can aid in this management problem.

In the article "Using Extranets in Fostering International Communities of Mathematical Inquiry," Lyn English and Donald Cudmore explore the possibilities of linking classrooms around the world into more diverse forms of learning communities. The example of an actual extranet that they describe demonstrates clearly the potential for classrooms of the future to be very different from what they are today.

Part 3 of the yearbook focuses on the mathematics curriculum. As the truism goes, "Nothing difficult is ever easy." Although some think this statement applies to learning mathematics, it certainly applies to deciding what should be in the mathematics curriculum of the twenty-first century. Johnny Lott and Terry Souhrada outline a century of debate over curriculum issues and argue that the dominant curriculum in secondary school mathematics of the twentieth century is not serving the needs of students entering the twenty-first century. An important lesson that emerges from their article is that changing the curriculum has never been easy. They conclude with a list of conditions that will be important if needed changes are to be made in the twenty-first century.

In the late twentieth century, the calls for change outlined by Lott and Souhrada led to curriculum development projects funded by the National Science Foundation. The article "The Impact of *Standards*-based Instructional Materials in Mathematics in the Classroom" discusses the materials generated by these projects as prototypes of standards-based curricula for the twenty-first century. In that article, Eric Robinson, Margaret Robinson, and John Maceli provide specific examples drawn from these materials to illustrate one of the central points of the NCTM *Standards:* mathematics is a sense-making activity. The article closes with a series of challenges that reflect a vision of a curriculum not unlike that of Devlin and Wallace in Part 1 of this volume.

Susan Forman and Lynn Steen present a vision of "functional mathematics" as a curriculum for the twenty-first century. In their article, titled "Beyond Eighth Grade: Functional Mathematics for Life and Work," they offer a somewhat detailed view of how the curriculum can be fashioned to meet the demands of citizenship and the technical, problem-solving needs of the workforce in the twenty-first century. Forman and Steen reiterate the arguments presented by others that the primary rationale for improving school mathematics is not international competitiveness but equity. And they propose functional mathematics as a way in which both equity and competitiveness can be achieved.

One of the significant additions to the school mathematics curriculum of the twentieth century was the study of statistics. The use of technology has greatly improved the opportunity for students to study this subject and has indeed changed the priorities in statistics education. Richard Scheaffer, in his article "Statistics for a New Century," argues that the grades K–12 statistics curriculum needs to change from being taught as a series of techniques to being taught as a process of thinking about the world. He discusses four areas that need to be stressed in the grades K–12 curriculum: exploratory data analysis, the fundamental concept of association, inferential reasoning, and principles of planning studies. Scheaffer closes the article with a specific set of curriculum recommendations for the elementary, middle, and high school grades.

Part 4 of the yearbook is titled "Improving Mathematical Learning Environments" because the classroom, whether face-to-face or virtual, will be the primary locus of "learning mathematics for a new century." It seems evident that environments that support students' reasoning and inquiry, that seek out and value every student's input, and that nurture equitable discourse communities will be places where students want to learn. In the prologue, Willoughby states, "If the student has learned to hate mathematics and the learning of mathematics, then I believe the schooling has done more harm than good." In Part 4 there are four papers that offer specific methods for achieving the kind of equitable environments needed in the twenty-first century for nurturing the student's willingness to learn mathematics and to use mathematics.

Kay McClain, Paul Cobb, and Koeno Gravemeijer demonstrate how starting inquiry with a significant and complex challenge, such as analyzing and comparing data distributions, can lead all students to constructing on their own the rudimentary tools of statistical data analysis so often taught in isolation and without motivation. Their article, "Supporting Students' Ways of Reasoning about Data," illustrates how technology can be flexibly adapted to supporting students' initial constructions. Every student is able to observe, explore, organize, and formulate hypotheses about distributions of data without first being taught the conventional tools.

Miriam Sherin, Edith Mendez, and David Louis share in "Talking about Math Talk" a simple method for building discourse communities in mathematics classrooms whether the classroom is a virtual classroom of the future or a face-to-face classroom of the present. Their three-prong method of *explain, build,* and *go beyond* allows students to learn how to engage in constructive dialogue on mathematical issues and at the same time learn to value the contributions of each student in the class.

Karen Fuson and her coauthors describe a comprehensive set of pedagogical principles intended to develop equity while encouraging high achievement for all students. Their set of six aspects, collectively referred to as a "Mathematics Equity Pedagogy," attempts to blend a variety of methods that have proved to be effective in twentieth-century classrooms. Although many articles in this yearbook mention the need for balanced approaches to the challenges facing mathematics education, this article illustrates with specifics how a balanced approach can be achieved.

In the article "Exploring Mathematics through Talking and Writing," David and Phyllis Whitin illustrate how children's literature can furnish child-friendly contexts needed for stimulating mathematical discussion and writing. They point out methods for evoking such discourse in ways that involve all students. For example, children are invited to be creative in changing a story to make a mathematical point. Every child gets to tell a story that incorporates her or his solution to the mathematical problem.

The yearbook closes with an epilogue by Kilpatrick and Silver titled "Unfinished Business: Challenges for Mathematics Educators in the Next Decades." Kilpatrick and Silver provide historical background for each of the main challenges described in their article. These challenges fall into the areas of ensuring mathematics for all, promoting students' understanding, maintaining balance in the curriculum, making assessment an opportunity for learning, and developing professional practice. It is fitting that this yearbook end by challenging researchers and practitioners in the twenty-first century to address the significant and unsolved problems that the twentieth century has handed down.

In the preparation of the 2000 Yearbook, a dedicated panel of mathematicians and mathematics educators reviewed sixty-five manuscripts. The diversity of vision that this panel brought to the task made the entire editorial process an incredible learning experience for each of us. For their assistance in the very difficult task of deciding which manuscripts to accept, I would like to extend my deepest thanks to the members of the Editorial Panel:

John Dossey	Illinois State University (retired), Normal, Illinois
Kimberley Girard	Glasgow High School, Glasgow, Montana
Carolyn Mahoney	California State University at San Marcos, San Marcos, California
Lee Zia	University of New Hampshire, Durham, New Hampshire

Special thanks are due to Frances Curcio, the general editor for the 1999 through 2001 Yearbooks. Through her diligence and hard work, Fran made the entire editorial process achieve its goals in a timely fashion. My heartfelt thanks are extended to the many talented individuals who submitted manuscripts for review. Whether their manuscripts were accepted or not, without the efforts and generosity of these individuals there would be no 2000 Yearbook. Finally, I would like to thank the editorial and production staff at the NCTM Headquarters Office in Reston, Virginia. Their dedication to the NCTM yearbook series is evident in the high quality of the final products.

Along with the members of the editorial and production staff, it is my hope that readers will find the 2000 Yearbook, like the other volumes in the series, to be a meeting ground of important ideas and that, over time, readers will fill its margins with thoughts and notes of their own.

Maurice J. Burke
2000 Yearbook Editor

1

Perspectives on Mathematics Education

Stephen S. Willoughby

MATHEMATICS has played an important role in the development of society from prehistoric times to the present. That role is more significant today than ever before and promises to become even more significant in the future. Thus, mathematics education is of great interest and is sometimes the subject of heated debate. Many of the arguments and practices that command attention today seem remarkably similar to those of the past.

WHERE HAVE WE BEEN?

Since entering elementary school in the 1930s, I have in one way or another been involved in mathematics education. Even though my understanding of its history, both before and after that time, is influenced by my personal experiences, my perceptions of the history of mathematics education may offer a perspective from which to view the potential of, and challenge for, mathematics education in the twenty-first century.

A comprehensive history of mathematics education from the ancient Mayans to the Cambridge Conference was published in 1970 (National Council of Teachers of Mathematics [NCTM] 1970). Many of the activities and discussions in that history and in a companion volume of readings from the history of mathematics education (Bidwell and Clason 1970) seem quite familiar today.

The amount of mathematics that is expected of students has dramatically increased over time. A look at the past three hundred years of the United States mathematics curriculum illustrates this growth. Harvard appointed its first professor of mathematics in 1726 and began to require arithmetic of entrants. By 1820, Harvard required algebra. Geometry was first required in 1844. In 1912, when David Eugene Smith reported to the International Commission on the Teaching of Mathematics (ICTM) (ICTM 1912), only half the high schools in the United States offered as much mathematics as a year of algebra, a year of geometry, and another half year of algebra. Fewer than 20 percent of high schools offered any higher level of mathematics. In 1950 very few students entering college had been exposed to calculus, even though many had studied four years of high school mathematics. By 1970 it was common for college freshmen to have had one or two semesters of high school calculus. Today, many colleges expect four years of mathematics, including a full year of calculus. Although the trend is likely to continue, much of what was taught in school mathematics in the nineteenth and early twentieth centuries, such as the pencil-and-paper square-root and cube-root algorithms and complex financial computations, has been eliminated from the curriculum because it is no longer believed to be important.

The compartmentalized mathematics curriculum of the United States (arithmetic, algebra, geometry, more algebra, and precalculus-calculus) is an artifact of the order in which colleges changed their entrance requirements. Major national recommendations regarding education since the 1894 Committee of Ten report (National Education Association 1894) have recommended greater integration of subjects. Excellent textbooks that integrate mathematics have been available in the United States at one time or another since the 1920s, but the inertia of teachers, parents, college entrance requirements, and the public have prevented as much integration as would have been desirable. Ideally, I believe, intuitive introductions to concepts from arithmetic, geometry, measurement, algebra, statistics, and probability should start at a very concrete level in kindergarten or before and should gradually become more abstract over the years, but they should always be integrated with one another in a way that makes sense to the learners.

The idea that studying mathematics makes one more logical has been under attack at least since Plato said, "I have hardly ever known a mathematician who was capable of reasoning" (Bartlett 1948). Thorndike and Woodworth (1901) appeared to have driven the final nail into the coffin at the beginning of the twentieth century when they showed that given the methods then used, there was little or no transfer from mathematical topics to other branches of knowledge and even very little transfer from one mathematical topic to another absent what they called "identical elements." Later William Brownell and others conducted experiments that showed that if students learn with understanding, they do in fact show considerable transfer to new topics (NCTM 1970).

In spite of the mounting evidence that mathematics learned first with understanding and then practiced for retention if necessary is more useful and transferable than mathematics learned only by rote (Hiebert 1999), much of the teaching of mathematics in the first half of the twentieth century was based on Thorndike's principles of strengthening bonds through stimulus-response activities (Thorndike 1922). Furthermore, much of the early research in mathematics education concentrated on establishing precisely which mathematical skills would be needed in various occupations and then developing curricula to train students to become proficient at the skills they might use in their chosen fields. Little consideration seemed to be given to the possibility that the skills needed to be successful in a given occupation might change over time or that people might work at several different occupations during their lifetimes.

The progressive education movement occurred concurrently with Thorndike's influence. The Progressive Education Association (PEA) was founded in 1919 on such principles as (1) children should have the freedom to develop naturally, (2) interest should be the motivation for all work, and (3) the teacher is a guide, not a taskmaster. Only mathematics that was interesting or useful to the child was to be studied. Newspapers and other publications were searched for examples of mathematics that people might need in their social lives. Mathematics that was not perceived as useful either in work or in social life was presumed not to be worth teaching. Children learned the mathematics that occurred as they pursued projects in which they were interested or that could be shown to be useful in the occupations they were likely to pursue. Teaching anything else was hard to defend. Even though the PEA agenda had a strong influence in only a small percentage of schools, the combination of the recommendations of Thorndike and of the PEA resulted in a substantial reduction in activities to strengthen the mathematical skills and comprehension of students.

World War II had a profound effect on people's ideas about the need for mathematics. Young men and women were unable to understand or perform the mathematics needed to operate weapons systems, maintain supplies, navigate ships and airplanes, and do other relatively simple tasks needed in the war effort. The military was forced to provide crash courses in mathematics and other subjects for recruits. Suddenly mathematics was important for prospective members of the armed forces. Unfortunately, even this apparent recognition of the need for better mathematics education tended to be applied specifically to males of European extraction.

The NCTM created the Commission on Post-War Plans, which issued three reports (Commission on Post-War Plans of the NCTM 1944, 1945, 1947) recommending that schools ensure mathematical literacy for all who could possibly achieve it (NCTM 1970, pp. 66–88). However, the commissioners apparently believed that the majority of students ought not to be expected to study anything so advanced as first-year algebra. Today their recommendations for the mathematically literate person seem less than ambitious.

As late as 1953 a professor teaching the graduate mathematics education course in which I was enrolled assured the class that any students with an IQ less than 110 could not learn algebra, so there was no point in allowing such students to take the course. Girls were discouraged from taking more than a minimal amount of mathematics. In my 1950 graduating high school class of 247 students, only 38 boys and not a single girl registered for the fourth year of high school mathematics. People from minority groups practically never majored in mathematics in college, and relatively few even took all the available mathematics in high school.

Since the 1950s people have gradually begun to realize not only the inequity in this sort of prejudice but also the damage to society that is caused by it. If we exclude the vast majority of our students from studying any substantial mathematics, we not only exclude them from many of the higher-paying occupations but also create a future in which we will not have enough scientists, engineers, accountants, computer scientists, and so on to maintain a viable economy and national defense. Many barriers to learning mathematics for the traditionally underrepresented groups have been eliminated, but many (such as malnutrition, the drug culture, an often unmotivating and poorly structured mathematics curriculum, underprepared teachers, society's low esteem for intellectual activities, and the attitudes of parents and teachers) still exist and must be removed. There is evidence that when taught properly, mathematics is accessible to all people. See, for example, Dilworth and Warren (1980), which showed that with appropriate materials and instruction, children of lower socioeconomic status could achieve higher raw scores in both concepts and skills than middle-class children and that all children could achieve considerably more than was generally expected.

After the Second World War, educators, scientists, and mathematicians continued to argue for more emphasis on mathematics in schools, but the public's acceptance of the need for better mathematics education seemed to dwindle. Then, on 4 October 1957, the Soviet Union launched the first Sputnik satellite. The National Defense Education Act was passed, and hundreds of millions of dollars became available through the National Science Foundation (which had been created in 1950 with a primary purpose of enhancing scientific research) for improving mathematics and science education.

"New math" was really born in 1951 with the creation of the University of Illinois Committee on School Mathematics (UICSM), but only after 1957 did the federal government provide substantial support for mathematics and science education. The most influential of the new-math projects was the School Mathematics Study Group (SMSG). Beyond the curriculum development projects, numerous institutes were conducted all over the country to improve the preparation of teachers of mathematics. Most of these institutes concentrated on content, but many also included some pedagogy. Many teachers who attended such institutes believe that the most valuable part of the new-math era was the teacher institutes that enhanced their mathematics knowledge.

Public controversy surrounding new math began almost immediately and continued even after most of the curricular modifications had disappeared or had become so commonplace that few people remembered their source. Critics claimed that new math was too abstract, was not related to the real world of either adults or children, used language that was unknown even to mathematicians and other educated adults, and was doing more harm than good in American education. Most of the criticism was supported by citing specific examples of bad mathematics or pedagogy, but the critics often generalized from a very small number of instances. Variation in new-math programs was as great as it had been in previous programs.

A significant event in the new-math era was the publication of the *Report of the Commission on Mathematics of the College Entrance Examination Board* (CEEB)(CEEB 1959), which encouraged the elimination of solid geometry and trigonometry as separate semester courses, a functional approach to trigonometry, a high school course in probability and statistics, and other reforms. Pronouncements of the CEEB greatly influence college preparatory courses, and ultimately all courses taught in school. This report was no exception.

In contrast to the CEEB report, the Cambridge Conference Report (Educational Services Incorporated 1963) was seen by many educators as a wildly optimistic proposal by mathematicians who had little connection with schools. The suggestion that by the year 2000 a good high school education should include two years of calculus followed by a semester each of abstract algebra and probability theory did seem a bit unrealistic. But in its emphasis on good mathematical reasoning, problem solving, reducing senseless drill, integrating topics, a spiral curriculum, improved preparation of teachers, and so on the Cambridge Conference seems prescient.

In the the early 1970s, two quite contradictory trends existed. A strong back-to-basics movement developed in reaction to the new math. This movement dominated the textbook market throughout the 1970s and early 1980s. But on a much smaller scale, new and innovative programs were being developed and tested both by some commercial publishers at their expense and by others with support from government and private foundations. These innovative programs could be considered precursors to the NCTM Standards in that they were based on very different principles from those of the back-to-basics movement and demonstrated that all children could learn a remarkable amount of mathematics, including higher-order-thinking skills, if appropriate instructional activities were used (Dilworth and Warren 1980).

During the back-to-basics era of the 1970s and 1980s many mathematics educators argued for a broader vision of mathematics. That broader vision was partially described in innovative textbooks, projects, and documents including the Euclid Conference of the National Institute of Education (NIE) (NIE 1975), a National Council of Supervisors of Mathematics (NCSM) position paper on basic skills (NCSM 1976), and *An Agenda for Action* (NCTM 1980) published by the National Council of Teachers of Mathematics.

Some of the innovative textbooks and projects of the 1970s operated on principles such as the following:

1. All students should become proficient with both the traditional basic skills and the higher-order-thinking skills of mathematics.
2. Mathematics should be developed from the learners' viewpoint and experience, sometimes using physical manipulatives.
3. Students should learn to understand the power and beauty of mathematics.
4. Graduates of a good mathematics program should enjoy using and learning mathematics.
5. Students should learn to communicate clearly and correctly about mathematical situations.

Because some of the innovations were untested, several of the innovators employed long-term longitudinal field tests to verify their hypotheses. Among the innovative programs were the Comprehensive School Mathematics Program (Central Midwestern Regional Educational Laboratory 1978), Algebra through Applications (Usiskin 1976), Mathematics for Everyone (Wirtz 1974), Mathematics Their Way (Baratta-Lorton 1976), and Real Math (Willoughby et al. 1976, 1981, 1985).

I had the good fortune to be involved in one of the projects to develop innovative textbooks. Between 1972 and 1985, I worked with a discerning, principled publisher (M. Blouke Carus) and three outstanding scholars and educators on the Real Math program. Although Real Math and other innovative programs of the time were not by big publishers' standards great financial successes, I believe they had a significant and positive effect on hundreds of thousands of students and on the course mathematics education took in the last quarter of the twentieth century. Both the Euclid Conference report and the NCSM paper (mentioned above) were directly influenced by some of the innovative projects.

In 1981, when I became president-elect of NCTM, I proposed several new initiatives, including the creation of NCTM standards for instructional materials based on the principles and experiences of recent innovative programs. Many NCTM members had serious reservations about the proposal both because they had a fear of a "national curriculum" and because the NCTM had serious financial problems at the time and could not afford a major new project. In spite of these reservations, in 1984 NCTM published a small set of standards for the selection and implementation of instructional materials. In the five years preceding 1981, the NCTM's net financial worth had dramatically decreased and its membership had fallen by twenty thousand to reach a nadir of fifty-seven thousand. By 1984 NCTM's net worth had increased and memberships were again on the rise. The new NCTM

Board of Directors and president were convinced that more good than harm would come from NCTM standards, and in 1986 the NCTM Commission on Standards for School Mathematics was formed. NCTM published several sets of standards from 1984 through 1995 (NCTM 1984, 1989, 1991, 1995).

The 1989 *Curriculum and Evaluation Standards for School Mathematics* had the most far-reaching effect of the standards published by NCTM. Some people interpreted the NCTM *Standards* as suggesting that most of the traditional basic skills be eliminated and that schools should concentrate all their efforts on teaching thinking and problem solving and other higher-order-thinking skills. This was never the intent of the NCTM *Standards,* but even before 1995, members of NCTM were convinced that a revised and consolidated set of standards was needed. *Principles and Standards for School Mathematics* (NCTM 2000), among other improvements, makes clearer (1) the need for both traditional basic skills and higher-order skills, (2) the equity goals embedded in the previous standards, (3) the need to incorporate technology appropriately in the curriculum, and (4) the NCTM vision that its Standards are to be seen as a work in progress that will continue to be revised in the future.

One of the trends that accelerated in the last decade of the twentieth century was the movement by state and local communities to adopt their own standards. Most of these efforts were substantially influenced by the NCTM *Standards*. However, there have been some unfortunate and unintended consequences of this movement. States have become more explicit about what must be taught at each grade level and sometimes publish their guidelines only months before textbooks must be submitted. This situation, combined with the great expense of submitting textbooks for large state adoptions, makes difficult or impossible the adoption of innovative programs that have been developed and field-tested over a long time with small textbook publishers. The danger inherent in this trend is that textbooks that are less coherent but easily adaptable to give a superficial appearance of meeting the large variety of state guidelines will have a greater likelihood of being adopted. Care must be taken in our well-intentioned pursuit of higher standards that we, as an educational community, do not diminish our ability to be innovative.

WHY IS MATHEMATICS EDUCATION IMPORTANT IN THE TWENTY-FIRST CENTURY?

The world is becoming more mathematical. Decisions are often made that could benefit from mathematical insight. However, the decision maker often does not think mathematically and frequently does not even realize that thinking mathematically might have helped. This lack of awareness, I believe, is a failure of both the mathematics we teach and the way we teach it.

The goal of teachers of mathematics should be to help people understand mathematics and to encourage them to believe it is both natural and enjoyable to continue using and learning mathematics. Therefore, it is essential that we teach in such a way that students will see mathematics as a sensible, natural, and enjoyable part of their environment. I believe that we have often failed to teach the appropriate mathematics and that the mathematics we have taught has been taught in such a way as to make students dislike both the mathematics and the learning of it, thus ensuring that even if they could use mathematics effectively, they would be unlikely to do so.

There have been times when learning mathematics seemed of little use. In the 1930s, mathematics educators spent inordinate amounts of time defending the very idea of teaching mathematics to schoolchildren. Even if we believe that the place of mathematics in the future curriculum is ensured, teachers should think about why we teach mathematics and what mathematics people will need in the future so we can do the job better.

People have often believed they were well educated in mathematics if they could perform operations in arithmetic, algebra, and trigonometry; had memorized many facts, theorems, and proofs; and could solve the kinds of problems posed in textbooks and commercial tests. Such people were never mathematically literate, although some could find gainful employment performing such tasks. Today, most of these traditional basic skills can be performed by inexpensive calculators and computers more quickly and reliably than by the most skilled humans, so an employer might buy a calculator rather than hire a person with only those skills. The decisions about when to use mathematics, what operations to perform, which facts and theorems are pertinent, and how to formulate and solve problems in mathematical terms and interpret the results, however, are human activities and are likely to remain so. No computer is likely to do these activities better than educated humans in the foreseeable future.

Many traditional basic skills, however, are still appropriate and necessary for efficient mathematical thought. If a person must be creative to discover that $7 + 8 = 15$ and similar facts, there will be little creativity left for serious problems. One of the challenges for mathematics education is to decide which of the traditional basic skills are still important and which can safely be removed from the curriculum. But we can be quite certain that the ability to apply many of the traditional basic skills automatically, without thought, will enhance our ability to think about more-complex matters. As Alfred North Whitehead said, "It is a profoundly erroneous truism ... that we should cultivate the habit of thinking of what we are doing. The precise opposite is the case. Civilization advances by extending the number of important operations which we can perform without thinking about them" (Newman 1956, p. 442).

To try to list and defend the mathematical skills that ought to be learned by every citizen is hazardous work at best. Important skills will be neglected and some people will believe too much emphasis has been placed on others. The following is, however, a list of some of the skills that I think are important for the general population. For people continuing in scientific or technical studies, much more mathematics is required.

I believe all people should have the following:

1. A solid understanding of the significance and use of numbers in counting, measuring, comparing, and ordering. This includes an understanding of positive and negative rational numbers and the relationship between fractions and decimals. It also involves the ability to recognize absurd mathematical statements.
2. Proficiency in the basic operations with whole numbers (with an emphasis on base-ten positional notation), decimals, fractions, and algebraic symbolism; an understanding of the meaning of rates, ratios, proportions, and percents; and an understanding of their relationship to one another and to the real world
3. The ability to avoid tedious algorithms if thinking can provide shortcuts
4. The ability to decide which operations, arguments, or other mathematical thinking (if any) are appropriate to a real situation
5. The ability to decide when approximations or estimates are appropriate and how to use them to solve real problems with sufficient precision
6. The ability to use technology intelligently, including the ability to decide which technology (if any) is most appropriate to a given situation and to recognize the limits of the technology used
7. The ability to collect, organize, and interpret data intelligently; to extrapolate or interpolate from that data; and to recognize unsound or misleading statistical procedures when confronted with them
8. An understanding of probability concepts, including the unpredictability of single events and the long-term stability of ratios of events. People must understand the difference between foolish gambling and intelligent risk taking.
9. An understanding of the role and importance of functions in modeling the real world and the ability to represent functions numerically, algebraically, graphically, and verbally
10. The ability to create and recognize valid and invalid arguments or proofs in mathematics and in other environments
11. A knowledge of various geometric relationships, such as the ratios of lengths, areas, and volumes in similar figures; the Pythagorean theorem; and so on and the ability to apply this knowledge and understanding to real problems

12. A knowledge of, and the ability to use, simple useful trigonometric information such as the shape of the graph of the sine function and physical situations it models, the facts that $\sin^2 + \cos^2 = 1$, $\sin 30° = 0.5$, $\tan 45° = 1$, and so on
13. An intuitive understanding of the foundations of calculus
14. The ability to recognize situations in which mathematical thinking is likely to be helpful, to formulate problems in mathematical terms, and to interpret the results of mathematical analysis so that others can understand them

None of the skills above will be of any use if the individual who has learned them has also learned to avoid mathematics whenever possible. If a person is unwilling to use the skills learned in school, the learning has been a waste of time. If the student has learned to hate mathematics and the learning of mathematics, then I believe the schooling has done more harm than good. Our overriding goal in the teaching of mathematics should be to provide experiences that are both instructive and sufficiently rewarding that the student will wish to go on using mathematics and learning mathematics even after outside compulsion has ceased. Using and learning mathematics should be lifetime pursuits involving either individual thought and study or formal or informal group activities. I believe society should provide more opportunity for continued learning of mathematics and other subjects.

For more than fifty years parents, teachers, and scholars have told me that mathematics should not be "fun and games." They seem to believe that the purpose of learning mathematics is to improve the soul by making the student endure the most unpleasant kinds of experiences. I believe that mathematics should have more to do with the mind than the soul and that fun and games are appropriate parts of a good education as long as the games and fun have a positive intellectual purpose. If a student plays a game in which addition facts are practiced and continues to play the game and thus continues to practice long after attention to a page of written addition exercises would have ceased, surely this is good. If the game is so designed as to create an environment in which the student is also encouraged to recognize, formulate, and solve real mathematical problems, that is even better. I do not mean to suggest that hard work is not expected. But the concentrated thought, perseverance, and effort involved in doing mathematics should be rewarded with exciting, worthwhile, and beautiful results.

Throughout the history of mathematics education, some people have seen an austere but compelling beauty in mathematics. Many of those people have become teachers of mathematics. They are often puzzled when their students do not see the same beauty. Teachers of mathematics are obliged, I believe, to do everything in their power to help their students experience the joy of mathematics. Some students will understand and appreciate the overwhelming beauty of the subject, some will value it for its usefulness, and some may see little more than the fun of playing mathematical games or the clever ways that mathematical thinking allows us to avoid pedestrian, dreary work. If a student has learned a substantial amount of mathematics and has experienced the joy of mathematics, the teacher has been successful. We all should learn mathematics because it is useful, beautiful, and fun.

WHAT IS THE ROLE OF TECHNOLOGY?

Changes in technology over the last half century have been so drastic that nobody could have predicted them. In 1952 I heard with disbelief that there was a machine on campus that could invert a 100 by 100 matrix in less than a day. The machine occupied an entire house loaded with air conditioners. Today, a machine that fits comfortably on my desk is far more efficient and useful. Even those who have lived through these changes have a difficult time appreciating their magnitude. As late as 1970 personal calculators cost hundreds of dollars, were as big as a large book, usually performed only the four basic arithmetic operations, and had to be plugged in to an electric outlet (or if battery operated, would require new, expensive batteries for every few hours of operation). The suggestion that a normal individual might some day own a computer would have seemed ludicrous to most people.

The radical change in available technology has made people more aware of the role of mathematics in our society. Some have suggested that because we have computers and calculators, people no longer need basic skills. Others say that we need all the skills we ever needed and computers can be used to teach those skills. Still others believe that technology must be withheld from students so they will learn the important basic skills. I believe that advanced technology requires us to know more and different mathematics. Computers can arrive at completely fallacious results if improper assumptions are used in programming them. Insurance agents, car dealers, banks, and others have been known to use specially programmed computers to beguile customers. Computers and even calculators can produce mathematical nonsense at a much faster rate than people have ever been able to do in the past.

Educated people must understand sufficient mathematics to be able to recognize absurd mathematical statements and understand that mathematical statements generated by computer are no more likely to be valid than those coming from a human being. We must teach people not to think of technology as an authority figure. Without mathematical understanding people of the twenty-first century will be victims of computer mysticism.

These considerations raise the question of using the computer as a teaching machine. The basic skills can be drilled into students by using a computer as a sort of electronic flash-card device. This use of computers gives students exactly the wrong message: "The computer is right. The human may not be. The computer will decide whether the human got it right." Computers must not be thought of as authority figures.

The Educational Testing Service released data in October 1998 regarding the National Assessment of Educational Progress. These data strongly suggested that when computers are used for drill and practice they actually appear to depress grades and certainly don't improve them. Conversely, when the computers were used "to more sophisticated ends, such as simulations and applications of math concepts to real-life problems," their use had a substantial positive effect (Holden 1998, p. 407).

Should calculators be withheld until students have become proficient with all the basic skills? The answer depends on how calculators are used (Hembree and Dessart 1986). If students regularly use a calculator to find 7×8, the calculators probably are not a positive influence, although we could argue that this corresponds to looking up the spelling of a word in the dictionary. But seen from the standpoint of accessibility, using a dictionary is more inconvenient than using a calculator, so the motivation to remember the calculated answer may be less.

After I introduce calculators to a class, I try to provide situations in which the calculator will obviously be an inconvenience as well as situations in which it will be useful. We might have a race in which half the class uses calculators but must push every key in the operation while the other half is allowed to use whatever procedures they like. Then the questions might be $7 \times 8 = ?$, $70 \times 8 = ?$, $700 \times 8 = ?$, $70 \times 80 = ?$, and so on, ending with $70\,000 \times 80\,000 = ?$ (which on some calculators will produce only an error message). With many different activities of this sort, students soon learn to think before grabbing a calculator and pushing keys. Similar activities requiring more or less sophistication can be constructed for all levels of mathematical development. We must teach people to use technology intelligently, including not using it at all when it is inappropriate. If calculators are withheld, students will not have an opportunity to learn to use them, or not use them, appropriately. Some highly instructive activities entirely within the grasp of normal students would require too much pedestrian calculation without some form of technology. Beyond that, computer-based and calculator-based laboratories allow the sophisticated study of functions that would be difficult or impossible without them.

Developments in technology have had, and will continue to have, a profound effect on the kind and amount of mathematics that will be needed to understand and function in the workplace and in the everyday world. To try to predict exactly what people will have to know would be foolhardy. However, we know that whatever mathematics children learn today will not be appropriate for all situations they are likely to face. A clear implication of the changes we see in technology is that people will have to go on learning after they have completed their traditional formal education. The successful user of mathematics in the future will be willing and able to continue using and learning mathematics throughout life.

WHERE SHALL WE GO?

In my lifetime and before, mathematics education has undergone numerous substantial changes but has remained remarkably stable in many respects. The arguments and practices of today seem to echo those of the past. Progress has been slow, but noticeable, toward greater equity, national standards, higher expectations for teachers and students, and other desirable goals. Some changes, however, such as the impact of technology, have been quite abrupt and can be expected to continue to be of great significance for mathematics education.

Although I know not where we will go, I have strongly held opinions about some of the things we should do to improve mathematics education:

- We must teach both traditional basic skills and higher-order skills. We can do so efficiently if we teach the basic skills through a guided-discovery approach and practice them in a rewarding problem-solving environment and emphasize thinking throughout.
- Students should be led to believe that they can figure out most mathematics whether they have forgotten a fact or never learned it. However, they should understand that they will be more efficient if the most-used facts and algorithms are automatic.
- When using new technology we should be certain that there is a clear pedagogical advantage or that by relieving students of tedious activity we are freeing their minds to think and be creative.
- Mathematics education must be seen as a lifetime activity. Facilities for continuing mathematics education should be available. Students should expect to continue using and learning mathematics and should enjoy doing so.
- Mathematics should be learned as an integrated whole, starting with activities that are concrete and intuitive for the learner. For older students, concepts that were theoretical may become concrete. Arithmetic is concrete to a student studying algebra at the age of thirteen or fourteen. A four-dimensional space may be concrete to a student studying complex variables.
- All students can and should learn mathematics and should be willing and able to use it effectively. Universal learning is both a matter of equity and a matter of national survival.
- Excellent teachers have various ways of helping students learn mathematics, and rather than prescribe particular methods, we should evaluate

the results: content knowledge, ability and willingness to use mathematics appropriately, and so on.
- Teacher preparation must be improved. Teachers should see the learning of mathematics and of better teaching methods as lifelong activities, and society must support such continued learning. Beyond that, society must make a more substantial effort to attract and keep the very best people as teachers of mathematics.

The world is changing. Technology is changing. Mathematics is changing. Mathematics education—and society's perception of, and support for, mathematics education—must change to meet the needs of the twenty-first century.

REFERENCES

Baratta-Lorton, Mary. *Mathematics Their Way.* Innovative Series. Menlo Park, Calif.: Addison-Wesley Publishing Co., 1976.

Bartlett, John. *Familiar Quotations,* 12th ed., s.v. "Plato: *The Republic,* Book VII, 531-E."

Bidwell, James K., and Robert G. Clason. *Readings in the History of Mathematics Education.* Washington, D.C.: National Council of Teachers of Mathematics, 1970.

Central Midwestern Regional Educational Laboratory. *Comprehensive School Mathematics Program: CSMP in Action.* Saint Louis, Mo.: Central Midwestern Regional Educational Laboratory, 1978.

College Entrance Examination Board. *Report of the Commission on Mathematics of the College Entrance Examination Board: Program for College Preparatory Mathematics.* Princeton, N.J.: Educational Testing Service, 1959.

Commission on Post-War Plans of the NCTM. "First Report of the Commission on Post-War Plans." *Mathematics Teacher* 37 (May 1944): 225–32.

———. "Guidance Report of the Commission on Post-War Plans." *Mathematics Teacher* 40 (July 1947): 315–39.

———. "Second Report of the Commission on Post-War Plans." *Mathematics Teacher* 38 (May 1945): 195–221.

Dilworth, Robert P., and Leonard M. Warren. *An Independent Investigation of Real Math: The Field-Testing and Learner Verification Studies.* La Salle, Ill.: Open Court Publishing Co., 1980.

Educational Services Incorporated. *Goals for School Mathematics: the Report of the ESI Cambridge Conference on School Mathematics.* Boston: Houghton Mifflin Co., 1963.

Hembree, Ray, and Donald J. Dessart. "Effects of Hand-Held Calculators in Precollege Mathematics Education: A Meta-Analysis." *Journal for Research in Mathematics Education* 17, no. 2 (1986): 83–99.

Hiebert, James. "Relationships between Research and the NCTM Standards." *Journal for Research in Mathematics Education* 30, no. 1 (1999): 3–19.

Holden, Constance. "Random Samples: Dubious Benefits for Early Computer Use." *Science,* 16 October 1998, p. 407.

International Commission on the Teaching of Mathematics. *Report of the American Commissioners of the International Commission on the Teaching of Mathematics.* USBE Bulletin 1912, no. 14. Washington, D.C.: Government Printing Office, 1912.

National Council of Supervisors of Mathematics. "Position Paper on Basic Mathematical Skills." Golden, Colo.: National Council of Supervisors of Mathematics, 1976.

National Council of Teachers of Mathematics. *An Agenda for Action: Recommendations for School Mathematics of the 1980s.* Reston, Va.: National Council of Teachers of Mathematics, 1980.

———. *Assessment Standards for School Mathematics.* Reston, Va.: National Council of Teachers of Mathematics, 1995.

———. *Curriculum and Evaluation Standards for School Mathematics.* Reston, Va.: National Council of Teachers of Mathematics, 1989

———. *A History of Mathematics Education in the United States and Canada.* Thirty-second Yearbook of the National Council of Teachers of Mathematics, edited by Phillip S. Jones and Arthur F. Coxford, Jr. Washington, D.C.: National Council of Teachers of Mathematics, 1970.

———. *Principles and Standards for School Mathematics.* Reston, Va.: National Council of Teachers of Mathematics, 2000.

———. *Professional Standards for Teaching Mathematics.* Reston, Va.: National Council of Teachers of Mathematics, 1991.

———. "Professional Standards for Selection and Implementation of Instructional Materials." Reston, Va.: National Council of Teachers of Mathematics, 1984.

National Education Association. *Report of the Committee of Ten on Secondary School Studies with the Reports of the Conferences Arranged by the Committees.* New York: American Book Co., 1894.

National Institute of Education. *Conference on Basic Mathematical Skills and Learning.* 2 vols. Washington, D.C.: National Institute of Education, U.S. Department of Health, Education, and Welfare, 1975.

Newman, James R., ed. *The World of Mathematics.* Vol. 2. New York: Simon & Schuster, 1956.

Thorndike, E. L., and R. S. Woodworth. "The Influence of Improvement in One Mental Function upon the Efficiency of Other Functions." *Psychological Review* 8 (1901): 247–61, 384–95, 553–64.

Thorndike, E. L. *The Psychology of Arithmetic.* New York: Macmillan, 1922.

Usiskin, Zalman. *Algebra through Applications.* Chicago: Department of Education, University of Chicago, 1976.

Willoughby, Stephen S., Carl Bereiter, Peter Hilton, and Joseph H. Rubinstein. *Real Math.* La Salle, Ill.: Open Court Publishing Co., 1976, 1981, 1985.

Wirtz, R. *Mathematics for Everyone.* Washington, D.C.: Curriculum Development Associates, 1974.

2

The Four Faces of Mathematics

Keith Devlin

MATHEMATICS presents four faces to the world:
1. Mathematics as computation, formal reasoning, and problem solving
2. Mathematics as a way of knowing
3. Mathematics as a creative medium
4. Applications of mathematics

For the most part, current grades K–12 education concentrates on the first face and makes some reference to the fourth face but pays little or no attention to the remaining two faces. In this article, I shall argue that as we enter the twenty-first century, mathematics education should display all four faces.

WHY EDUCATE?

What is the purpose of a school education? In times past, it might have been possible to answer that at least one function was to provide young people with (at least some of) the knowledge and skills required for different occupations. Even if such a view was once defensible, it surely does not apply today, when most occupations require very specialized knowledge and skills and where the pace of change is such that no one can predict more than a few years in advance what the most important knowledge and skills are going to be—save for the crucial importance of one particular ability: the ability to adapt to changing circumstances and to acquire rapidly a new set of skills.

A much better answer, in my view, is that one purpose of a school education is to develop in young people *the ability to acquire specialized knowledge and skills* as and when they need them during the course of their working lives.

I think the second answer is much better. But I also think that preparation for work is just one part of the real purpose of a school education. That real purpose, as I see it, is to pass on the main elements of our culture and to prepare young people to lead full and active lives, playing their full roles as citizens. This goal is very broad and ambitious, probably not one that a poor or developing nation can afford. But it is, I suggest, the only goal appropriate for any society, such as ours, that has become sufficiently affluent for most people to have considerable choice about how they want to live their lives.

The question I want to investigate in this essay is, What are the implications of this view of education when it comes to mathematics?

Why Mathematics?

Why does mathematics form a part of the education we provide for our young? Let me start my answer, as I did above for education in general, by concentrating on the utilitarian aspect of mathematics: How much mathematics does a typical young person *need* to learn to be adequately prepared for his or her subsequent occupations?

The simple answer is, much less than is popularly assumed. Few citizens in modern society ever need or make real use of any appreciable knowledge of, or skill in, mathematics. What mathematics they do need and use they have already met by the time they are twelve years old. (It is interesting that prior to that age, the majority of children declare that they "like" mathematics.)

The continuation of modern society, however, requires a steady supply of a small number of individuals having considerable training in mathematics. Just as in the industrial age we burned fossil fuels to drive the engines of production, so in today's information age the principal fuel we burn is mathematics. Whatever else we do with mathematics education, in order that the future supply of mathematicians does not dry up, we must ensure that all high school and university students are made aware of the nature and importance of mathematics so that those who find they have an interest in the subject, and an aptitude for it, can choose to study it in depth.

The two observations above have an obvious consequence for grades K–12 mathematics education: a major goal should be to create an awareness of the nature of mathematics and the role it plays in contemporary society. Grades K–12 mathematics should therefore be taught much more like history or geography or English literature—not as a utilitarian toolbox but as a part of human culture. Remember that the goal I am advocating is to produce an educated citizen, not a poor imitation of a twenty-dollar calculator. An educated citizen should be able to answer the following two questions about mathematics:

- What is mathematics?
- Where and how is mathematics used?

Few people can answer either question completely. In a moment I shall give my own answers to these two questions—answers that I believe an adequate mathematical education should equip everyone to provide. In the meantime, let me observe that a shift away from a largely procedure-oriented approach to mathematics to one that encompasses mathematics as a part of human culture—as a broad body of human knowledge and a "way of knowing"—is very much in keeping with the overall philosophy of the purpose of school education I stated at the beginning of the article. And yet now I have reached the same conclusion from the more obviously utilitarian standpoint of preparation for future careers.

In order to give our students a broad view of mathematics as a part of human culture that we need to address all four faces of mathematics. How do we go about doing so?

The Familiar Face

The first face of mathematics (computation, formal reasoning, and problem solving) is the one most familiar to the majority of people. This is the face we generally concentrate on when we think of mathematics education. Although it is so familiar that I will not dwell on it in this article, I am in no way advocating that we abandon this aspect of mathematics. Although the pocket calculator and the computer have made it unnecessary to spend large amounts of time developing skill in mental arithmetic, basic arithmetic and a good sense of number are surely invaluable skills for everyone to have in today's world. Moreover, learning mathematics seems to provide a general benefit in developing the mind that cannot be obtained any other way. For instance, even though few people ever make explicit use of high school algebra, it has been shown (see, e.g., U.S. Department of Education [1997]) that not only do students who complete a rigorous high school course in algebra or geometry more frequently gain entry into a college or university, but they perform much better once they are there whatever they choose to study. In short, completing a rigorous course in mathematics—it is not even necessary that the student do well in such a course—appears to be an excellent means of sharpening the mind and developing skills in problem solving and analytic thinking.

Despite the benefits, however, if we continue to concentrate largely on the first face of mathematics and ignore the remaining three faces, we not only fail our students—our future citizens—we also do a great disservice to one of humankind's most towering cultural achievements.

Mathematics as a Way of Knowing

Thought of as a human activity, mathematics is a particular way of knowing, a way of understanding different aspects of the world we live in. Mathematics is not the only way of knowing by any means: biology, chemistry, physics, psychology, sociology, linguistics, poetry, painting, sculpting, playwriting, and novel writing are just some of a great many other ways of knowing. Mathematics, however, is a way of knowing that has proved to be inordinately successful. Today, scarcely any aspect of our lives is not affected, often in a fundamental and far-reaching way, by the products of mathematics. When you think of the technological and communications infrastructure that undergirds our lives, you realize that we are in fact living in a "mathematical universe."

Each of the different ways of knowing has its own particular characteristics. What exactly is the way of knowing we call *mathematics?* Most people think that mathematics is merely a collection of rules for manipulating numbers, but they are wrong. The manipulation of numbers is just one very small part of mathematics. The simplest, most accurate description of mathematics I know is this: Mathematics is the science of patterns. The mathematician looks at a certain aspect of the world and strips away the complexity, leaving an underlying skeleton. Looking at different aspects of the world in this way leads to different branches of mathematics, which focus on different kinds of patterns. For instance:

- Arithmetic and number theory study the patterns of number and counting.
- Geometry studies the patterns of shape.
- Calculus allows us to handle patterns of motion.
- Logic studies patterns of reasoning.
- Probability theory deals with patterns of chance.
- Topology studies patterns of closeness and position.
- Projective geometry arises from a study of the patterns that enable us to perceive depth in a two-dimensional picture.
- Group theory results from a study of the patterns of symmetry.

It would of course be impractical to try to teach students how to do each of these different kinds of mathematics (to say nothing of all the branches I have not listed), and I am not advocating doing so. Rather, I think it is important to make our students aware of the existence of these many different branches of the subject and of what makes them all *mathematics.*

How do we go about creating that awareness? My answer is that we do so by describing some of the many applications of mathematics. In so doing, we show our students that mathematics works by *making the invisible visible.* We show them that by giving us a means to "see" (and hence to understand) things that would otherwise be invisible, mathematics demonstrates that it is one of the most amazing constructions of the human mind, a powerful testimony to human ingenuity and intellectual creativity. The following are just a few examples of the kind of thing I have in mind.

Without mathematics, there is no way to understand what keeps a jumbo jet in the air. As we all know, large metal objects don't stay above the ground without support. But we can't see anything holding up a jet aircraft. It takes mathematics to help us "see" what keeps an airplane aloft. In this instance, what lets us "see" the invisible is an equation discovered by the mathematician Daniel Bernoulli early in the eighteenth century.

What causes objects other than aircraft to fall to the ground when we release them? "Gravity," you answer. But you are simply giving the phenomenon a name. We might as well call it magic. It's still invisible. To understand it, we have to "see" it. That's exactly what Newton did with his equations of motion and mechanics in the seventeenth century. Newton's mathematics enabled us to "see" the invisible forces that keep the earth rotating around the sun and cause an apple to fall from a tree onto the ground.

Both Bernoulli's equation and Newton's equations use calculus. Calculus works by making visible the infinitesimally small. That's another example of making the invisible visible. Here's another:

Two thousand years before we could send spacecraft into outer space to take pictures of our planet, the Greek mathematician Eratosthenes used mathematics to show that the earth was round. Indeed, he calculated its diameter, and hence its curvature, with 99 percent accuracy. Today, we may be close to extending Eratosthenes' feat and discovering whether the universe is curved. Using mathematics and powerful telescopes, we can "see" into the outer reaches of the universe. According to some astronomers, we will soon see far enough to be able to detect and measure any curvature in space.

If we can calculate the curvature of space, then we can use mathematics to see into the future to the day the universe comes to an end. Using mathematics in conjunction with scientific theories, we have already been able to see into the distant past, making visible the otherwise invisible moments when the universe was first created in what is called the *big bang*.

Back on earth, what makes the pictures and sound of a football game miraculously appear on a television screen on the other side of town? One answer is that the pictures and sound are transmitted by radio waves—a special case of what we call *electromagnetic radiation*. But as with gravity, the term just gives the phenomenon a name; it doesn't help us "see" it. In order to "see" radio waves, we have to use mathematics. Maxwell's equations, discovered in the last century, make visible to us the otherwise invisible radio waves.

Mathematics has been used to describe human patterns, as well:

- Aristotle used mathematics to try to "see" the invisible patterns of sound that we recognize as music.

- Aristotle also used mathematics to try to describe the invisible structure of a dramatic performance.

- In the 1950s, the linguist Noam Chomsky used mathematics to "see" and describe the invisible, abstract patterns of words that we recognize as a grammatical sentence. He thereby turned linguistics from a fairly obscure branch of anthropology into a thriving mathematical science.

Finally, using mathematics, we are able to look into the future:
- Probability theory and mathematical statistics help us predict the outcomes of elections, often with remarkable accuracy.
- We use calculus to predict tomorrow's weather.
- Market analysts use various mathematical theories to try to predict the future behavior of the stock market.
- Insurance companies use statistics and probability theory to predict the likelihood of an accident during the coming year, and they set their premiums accordingly.

Mathematics allows us to make visible another invisible when it helps us predict the future. In that situation our mathematical vision is not perfect; our predictions are sometimes wrong. But without mathematics, we cannot see even poorly.

Let me stress once again that I am not advocating that we teach our students how to perform the mathematics involved in the applications mentioned above. Rather, along with the mathematics that we require our students to carry out, we should also *describe* some of the many other (perhaps more advanced) branches of the subject and give examples of the different ways they can be applied. Just as it is not necessary to know how to build or repair a car in order to take a tour in the country, so too it is not necessary to know how to do mathematics in order to understand how it is used. This analogy works just as well for the third face of mathematics, to which I turn next.

Mathematics as a Creative Medium

Few people are aware of the breadth of modern mathematics. Even fewer people realize that mathematics can also be used as a creative medium, in much the same way that a sculptor uses stone, a painter uses paint and canvas, or a novelist uses language. Used as a creative medium, mathematics again makes the invisible visible. In this instance, it takes the creative ideas produced in our minds, which are invisible to others, and makes them accessible to public perception, so others can share in them and experience our ideas. Mathematics as a creative medium is the third face of mathematics.

Arguably the first major creative use of mathematics occurred in the Renaissance, when artists discovered how to show depth in a two-dimensional painting. Artists refer to the trick as *the rules of perspective;* mathematicians call it *projective geometry*. Whatever the terminology, the underlying idea is to discover and use a "geometry"—the geometry of vision.

Similar to the artists of the Renaissance, present day artists have learned to use a geometry of light (so-called *ray tracing*) to produce realistic-looking computer graphics for the movie industry, images with surface texture, highlighting, and shadow.

More generally, much of the digital special-effects work in today's movie industry results from the use of mathematics as a creative medium. The special-effects artists use computers to create and manipulate mathematical descriptions of images that become visible only at the end of the process, when the massive arrays of numbers generated and stored in the computer are turned into colored pixels on a screen or a film.

Long before we had computers, the British writer Edwin A. Abbott, in his delightful novella *Flatland* (Abbott 1991), used the geometry of two and three dimensions as the vehicle for an insightful satirical commentary on the social mores of Victorian England. In the present era, the playwright Tom Stoppard has used mathematics as a vehicle for commenting on society in a number of his plays, of which *Arcadia* (Stoppard 1996) is a prime example.

Both Abbott and Stoppard provide examples of creative uses of mathematics not by mathematicians but by artists—artists who might well declare themselves ignorant of mathematics (although in most instances their ignorance is not so much of mathematics but of the fact that what they are doing has mathematical aspects). In another example, the discovery of non-Euclidean geometries and the investigations of four-dimensional geometry in the nineteenth century inspired many artists to explore and experiment with the nature of space and of dimension. A notable artistic development of this kind, with quite obvious mathematical roots, was the cubist movement in painting, led by Pablo Picasso and others. The Dutch artist M. C. Escher was another who tried to express different geometries in his paintings and etchings. Escher did in fact study mathematics, and he sometimes made the mathematics in his work fairly explicit.

More recently, the artist Tony Robbin (see Devlin [1998]) has spent a large part of his career trying to depict four-dimensional space on a two-dimensional canvas—a sort of "super perspective," if you will. According to Robbin, one of the main functions of art is to reflect on, comment on, and thereby help us understand various aspects of life. He sees his own work in exploring higher-dimensional spaces as a way to visualize and understand the multidimensional complexities of life in multiracial and multicultural societies.

Virtual reality art is another domain of modern art that is heavily dependent on mathematics. Artists such as Marcus Novak (see Devlin [1998]) use mathematics to create "immersive experiences" in which the user dons a stereoscopic and stereophonic helmet and wears a special pointing glove to "step inside and explore" the artistic world that the artist has created using mathematics, a world that can be multidimensional or structured according to a geometry very different from the one we are familiar with from our everyday world.

Of course, the use of mathematics as a creative medium is an application of mathematics (and so too is its use as a way of knowing). So we have already encountered the fourth face of mathematics: applications of mathematics. But I want to consider the application of mathematics separately, as a face in its own right.

Applications

Mathematics education must include applications. In order to live a full life, everyone needs to have an awareness of what goes into making that life possible. In particular, modern society depends on the many applications of mathematics that have been developed over the centuries, especially over the past half century. When we travel by car, train, or airplane, we enter a world that depends on mathematics. When we converse on the telephone, or attend a major sporting event, we are enjoying the products of mathematics. When we listen to music on a compact disc or log on to the Internet, we are using the products of mathematics. When we go into the hospital or take out insurance, we are depending on mathematics. As educators, we owe it to our students to make them aware of the scope, the depth, and the profound impact of the applications of mathematics in today's world.

Of course, much of the mathematics that lies behind our everyday world is advanced and highly specialized, and there is certainly no need for all but a small number of experts to understand that mathematics. Consequently, most classroom discussions of the applications of mathematics will have to be just that—discussions. But not all applications of mathematics are inaccessible. Moreover, with improving computer technology, it is getting easier for us to have our students actually carry out some well-chosen applications of mathematics. Thus, we are not restricted solely to talking about applications; we can have our students carry out some applications—an activity that should definitely be included along with the "tour" of the other applications I am advocating.

APPRECIATING MATHEMATICS

By changing our mathematics education system radically so that the primary goal for the vast majority of students is to create an awareness of the what, the how, and the why of mathematics rather than develop skills that only a tiny minority of the students will ever use, we will achieve two important goals:

1. That tomorrow's citizens appreciate the pervasive role played by one of the main formative influences on the culture in which they live.

2. That the individuals who turn out to have an interest in, and a talent for, advanced mathematics be exposed to the true nature and extent of the subject at an early age and as a result have an opportunity to pursue their interest to the eventual benefit of both themselves and society as a whole.

The justification for goal 1 is simply this: A human being is the richer for having a greater understanding of the nature of her or his life. The more ways we have to know our world and ourselves, the richer our lives are.

Regarding goal 2, success at high school mathematics, including calculus, is not always a good predictor of later success in mathematics. Mathematics through calculus is largely (though not exclusively) algorithmic: A successful strategy adopted by many students is simply to learn various rules and procedures and know when and how to apply them. In contrast, much (though not all) college-level mathematics beyond calculus is highly creative, requiring original thought and the ability to see things in novel ways. Since the creative mathematician often needs to apply rules and use algorithmic thinking, many successful mathematicians have indeed excelled in their high school mathematics classes. But many students who shone at high school mathematics find that they struggle with, and eventually give up, the subject in college, when they discover that algorithmic ability alone is not enough. And the fact that some of the very best professional mathematicians did poorly at high school mathematics but by some fluke were drawn to the discipline later in life suggests that our present system of school mathematics education probably turns off a significant number of students who have the talent for later mathematical greatness.

Quantitative Literacy

The focus of this article has been on mathematics education, primarily for the vast majority who do not make explicit use of mathematics in their lives. I have largely ignored what is often referred to as *quantitative literacy*. It is important to recognize that mathematics and quantitative literacy are not the same. Roughly speaking, quantitative literacy—sometimes called *numeracy*—comprises a reasonable sense of number, including the ability to estimate orders of magnitude within a certain range, the ability to understand numerical data, the ability to read a chart or a graph, and the ability to follow an argument based on numerical or statistical evidence. (See, e.g., Steen [1997].)

Since people often confuse quantitative literacy with mathematics, I shall address the former, even though it is not the focus of this article.

In today's society, numeracy is a fundamental life skill, on a par with literacy. In consequence, just as literacy is every teacher's responsibility—to be developed at all times in every lesson, not just in the English class—so too is numeracy every teacher's responsibility. It is as much the responsibility of the teacher of history, of home economics, or of physical education as it is the responsibility of the teacher of mathematics or science. Regarding basic quantitative skills as somehow separate from basic language skills sends quite the wrong message to our students. Confusing quantitative literacy with mathematics simply confounds the problem.

According to some estimates, fewer than 10 percent of the adult population of the United States is quantitatively literate. Such figures are difficult to interpret. One reason is that quantitative literacy has no fixed standard against which it can be measured by a test. What seems beyond doubt, however, is that, as a nation the United States is most definitely not quantitatively literate. I believe that this deficiency is a result of our (1) not regarding quantitative literacy as a basic responsibility of *all* teachers, on a par with basic language skills, (2) confusing quantitative literacy with mathematics, and (3) chasing the goal of achieving widespread proficiency in parts of *mathematics,* which for all but a small minority is simply unattainable (and, contrary to an oft expressed opinion that the decline in mathematics skills is a recent phenomenon, was probably not attained in the past, either).

Many more people appreciate music than can play a musical instrument. Many more people can enjoy a good novel or play than could write one. Many more people enjoy the benefits of driving an automobile than have the knowledge or skill to repair one. Similarly, we should recognize that it is possible to help people appreciate mathematics without forcing them (in vain) to try to achieve a working skill in the discipline. If we recognize that quantitative literacy and mathematics are different, if we accept that quantitative literacy is everybody's responsibility, and if we teach mathematics with the goal of developing an awareness of, and an appreciation for, its nature, extent, and relevance to modern life, then I see no reason why we cannot make a dramatic improvement in both the overall level of quantitative literacy and the presently poor level of general mathematical (and scientific) literacy in the population at large.

How Do We Do It?

Providing students with the broad overview of mathematics I am advocating will require a major change in the way we prepare high school mathematics teachers. At the same time, ensuring that we do not in the process lose the equally important goal of quantitative literacy requires that we change the way we prepare *all* high school teachers, not just the mathematics and science teachers.

Providing a broad overview of mathematics to today's students will require the imaginative use of all available media, in particular video and interactive computer technologies, as well as the preparation of first-rate printed material. I am avoiding using the word *textbooks,* since I do not believe that textbooks are suitable for the kind of instruction I am suggesting. Far better are high-quality expository books of the kind usually referred to as *popular science books*. I list a number of such books at the end of this article.

In many ways, what I am advocating does not require the development of a radical new way of teaching. Teachers of history, geography, economics, and psychology know how to teach subjects that have both descriptive, factual content and procedural aspects. Indeed, in those disciplines, the teaching is generally more factual and descriptive than procedural. My suggestion really amounts to adopting a similar style of teaching for mathematics, though perhaps with a different balance between the descriptive and the procedural elements.

Of course, mathematics is too extensive to expect any teacher to be able to master it all. All of mathematics, however, consists of variations on the same theme: the identification, abstraction, study, and application of patterns, using the mental tools of logical reasoning. A teacher who has mastered one area of mathematics will have little difficulty guiding students on a voyage of discovery in any other branch, provided that the experts in those other areas produce sufficiently accessible expository materials. An example of the kinds of material I have in mind is the 1998 Public Broadcasting Service production *Life by the Numbers*. Produced by WQED Television in Pittsburgh, this seven-part television series, also available on videocassette, showed the enormous extent of modern mathematics and the role it plays in all aspects of our lives.

In the end, however, education is about people. No matter how much money we spend on stimulating books, glossy television programs, and fancy computer products, we will not make any real progress unless we channel far greater funds into teacher preparation than we have in the recent past. That issue is societal. I hope it is not too idealistic a dream to hope that by restructuring grades K–12 mathematics education to provide a broad overview of the nature and breadth of mathematics and the role it plays in modern society, we will produce a new generation of public leaders who are themselves sufficiently appreciative of mathematics to recognize the importance of developing an adequate supply of well-prepared mathematics teachers, equipped with the resources necessary to give our young people the mathematics education they need and deserve. As I observed earlier, mathematics is the fuel that drives the information age. And education is the way we manufacture that fuel.

References

Abbott, Edwin A. *Flatland: A Romance in Many Dimensions.* Princeton, N.J.: Princeton University Press, 1991.

Devlin, Keith. *Life by the Numbers.* New York: John Wiley & Sons, 1998.

Steen, Lynn Arthur, ed. *Why Numbers Count: Quantitative Literacy for Tomorrow's America.* New York: College Entrance Examination Board, 1997.

Stoppard, Tom. *Arcadia.* Boston: Faber & Faber, 1996.

U.S. Department of Education. *Mathematics Equals Opportunity: A White Paper Prepared for US Secretary of Education Richard W. Riley, October 20, 1997.* Washington, D.C.: U.S. Department of Education, 1997. Available at www.ed.gov/pubs/math.

Additional Reading

Devlin, Keith. *The Language of Mathematics: Making the Invisible Visible.* New York: W. H. Freeman & Co., 1998.

———. *Mathematics: The New Golden Age.* 2nd. ed. New York: Columbia University Press, 1999.

Dunham, William. *The Mathematical Universe: An Alphabetical Journey through the Great Proofs, Problems, and Personalities.* New York: John Wiley & Sons, 1994.

Peterson, Ivars. *Islands of Truth: A Mathematical Mystery Cruise.* New York: W. H. Freeman & Co., 1990.

———. *The Mathematical Tourist: Snapshots of Modern Mathematics.* New York: W. H. Freeman & Co., 1988.

Steen, Lynn Arthur, ed. *On the Shoulders of Giants: New Approaches to Numeracy.* Washington, D.C.: National Academy Press, 1990.

Stewart, Ian. *The Magical Maze: Seeing the World through Mathematical Eyes.* New York: John Wiley & Sons, 1998.

———. *Nature's Numbers: The Unreal Reality of Mathematical Imagination.* New York: Basic Books, 1995.

3

The Many Roads to Numeracy

Dorothy Wallace

IN THE wake of performance standards for precollege mathematics has come a hue and cry for an ill-defined but widely applauded goal of "quantitative literacy" at the college level. Texts have sprung up like mushrooms after a rain, designed to meet this goal head-on with hundreds of pages of exercises that vary from remedial to indefensible. One text has topics ranging from the algebra of the logarithm to a thin and somewhat inaccurate exposition of Gödel's incompleteness theorem. As a mathematician, as a parent, as a product of a mostly public education, one has to wonder where all this is heading. Few arguments have been offered for the necessity of an entire population to understand logarithms that could not equally well be offered on behalf of Kurt Gödel and his famous theorem. So the issue of where to draw the line representing quantitative literacy can be debated at length, and in the absence of either cultural consensus or solid reason, the debate can become quite shrill.

The reasons behind the push toward standards for "quantitative literacy," whatever it is, are good ones. The economy and the development of the country depend on the presence of a certain proportion of technically savvy citizens. The proper working of democracy requires everyone to bring thought and opinion to bear on issues that are technical and often quite mathematical, even though it is hard to foresee exactly what these issues will be. So without knowing just what sort of mathematical expertise will prove valuable next year, the only recourse for those setting standards is to prescribe a full course—all the way from logarithms to incompleteness—because one just never knows what might be needed. That response may be logical, but it is almost certainly unfeasible. Even mathematicians aren't usually omnivorous enough to digest everything composing such a diet.

Any alternative to the prospect of universal standards must be manageable, must address the needs of the country, and must also address the needs of the individual student. But before an alternative can be offered, we must first revisit the basic question that such a standard addresses. Our educational system is based on a compromise between individual fulfillment and societal demands, exactly as it should be. So the question at the heart of the quantitative literacy debate, "What parts of mathematics are so basic that everyone ought to know them?" has to be approached by at least these two paths, which are very different indeed.

QUANTITATIVE LITERACY AND INDIVIDUAL FULFILLMENT

Let us look first at how the question pertains to an individual, in this example, a college student. Since 1994, Dartmouth College has made itself the site of an experiment that we (quite unoriginally) call the Math across the Curriculum project. One of seven projects funded by the National Science Foundation as part of its Mathematics and Its Applications throughout the Curriculum initiative, the Dartmouth project sought to integrate mathematics with a variety of disciplines throughout the institution. This meant that a cadre of mathematicians made themselves nuisances in various corners of the college by asking colleagues in many departments, "How can mathematics enrich, broaden, or serve your students?" Once humanities professors got over their initial surprise at even being asked such a question, they offered several insightful answers. They believed that the right sort of mathematics education had the potential to give a student confidence to tackle certain aspects of our culture (like technology) that were relevant to their chosen discipline of study. Students might become more aware of the multiplicity of answers and approaches to complex problems as well as develop a healthy skepticism toward what one faculty member called "the authority of numbers." Students comfortable with mathematics would be able to ask new kinds of questions and follow new lines of inquiry in any field, would be better problem solvers in general, and would be more accustomed to the abstraction and building of conceptual structure that accompanies much intellectual work. Some believed that mathematics furnished a unique window on the workings of the human mind, thus offering insight useful to the humanist. One humanist observed that in philosophy or literature, there was a collection of works widely accepted as "greatest hits," so important to the formation of our culture that all students should experience at least some of them. This professor went on to observe that mathematics and science must have some developments of equal importance, currently completely absent from the cultural curriculum and whose absence represents a great loss to students. These faculty members recommended that when educating humanities students, mathematicians should let go somewhat of the students' mastery of technique, focusing instead on connecting ideas and comparing the processes of thinking that distinguish mathematics and the humanities. These recommendations fit very well with the faculty's own professed goals for their students: learning to analyze and abstract, learning both to separate and to integrate emotional, aesthetic, and academic truth, and understanding the context for cultural development.

The result of this conversation was both a revision of existing courses to include a mathematical component and the creation of a collection of new, interdisciplinary courses, many of which involve mathematics in the humanities or arts. Topics include "The Mathematics and Philosophy of Infinity," which explores the ideas of Newton, Cantor, and others about the infinitely large and the infinitesimally small; "Music and Computers," which addresses questions ranging from signal processing to the composition of musical pieces; "Renaissance Astronomy and the New Universe," which treats the Copernican revolution from literary, scientific, and mathematical viewpoints; and many other courses as well. The mathematics in these courses ranges from elementary geometry to relativity, representing a real effort to bring to nonscience students the mathematics that was once the private reserve of those majoring in mathematics.

The students in these new courses, although enrolled for varying reasons, often share a distaste for traditional sorts of mathematics courses and even a slight fear of mathematics. What attracted them to the courses was the appeal made to their love of art, literature, or music. These courses, without stinting on content, directly addressed the preferences and inclinations of the individual. In one instance we introduced undergraduates to group theory as it applies to design problems and repeat patterns and witnessed a real willingness of these nonmathematical students to think like mathematicians in the service of design. Students in the "Pattern" course spent about half their time working on design problems, many of which depended on principles from the classification theory of the two-way repeat pattern. The other half of the time they thought about simple questions in group theory, many of which arose directly out of their art. As one student wrote, "It was the first time I ever applied math to life. It's neat to be able to look at it and apply it to other things." Relevance, like beauty, is in the eye of the beholder.

Other comments offered by students indicated that, generally speaking, when one of these interdisciplinary courses succeeded in appealing to an individual, it was because it addressed one of the goals stated by the faculty we had interviewed. For example, a student of mathematics and music remarked, "What I found ultimately was that the math component of music is what ties it to us as humans." So, as far as intellectual growth of the individual is concerned, the observations of our colleagues in humanities seem to be accurately reflected in the perceptions of our students. We can, therefore, draw a strong conclusion to the first part of our question about what mathematics students need to know. We are quite sure that when the role of education is to provide an intellectually fulfilling experience for the humanities undergraduate, a good strategy is to offer a wide variety of interdisciplinary courses in which students study a specific topic in depth, at the same time targeting the goals of connecting ideas in and out of mathematics and illuminating the thought processes of different disciplines.

QUANTITATIVE LITERACY AND SOCIETAL DEMANDS

Our second approach to the question of "What mathematics?" demands that we ascend a much steeper and rockier path. We must begin by seeing our society and culture for what it is, that is, more than merely an aggregate of individuals. Our culture will not succeed or fail, grow or diminish, because of the behavior or qualifications of any particular individual in it. We succeed because, as a whole, we are able to adapt to the changing circumstances of environment and history. If, as a whole society, we are not able to cope with advances in technology, greater speed of communication, new ideas and moral challenges, then we will fail. Perhaps the economy won't prosper, perhaps development will lag, perhaps the social structure will suffer, because we have created two classes of citizen. Adaptability is the key.

The ecological parallel offers a useful perspective on the problem. Like an ecosystem, neither society nor the economy will necessarily falter if some individuals drop out of it, if some entire vocations disappear, if new ones appear. What matters is its ongoing adaptability. Seen from this viewpoint, the prospect of universal standards in any subject looks very different. It is possible to argue that, for adaptability's sake, it really is necessary for nearly everybody to be able to read, do basic arithmetic, and even understand what a function is and how it can express information. Such arguments have been central to the development of the National Council of Teachers of Mathematics (NCTM) Standards and other standards for our grades K–12 educational system, which must address the basic needs of all our children. However, it is nearly impossible to argue that almost everybody must understand how to compute with logarithms. In fact, the ecological metaphor makes it clear that setting such a standard could well be counterproductive. Suppose one could, for example, remove all the animals from the forest who couldn't manage a minimum speed of locomotion on the ground. Would the resulting forest be more adaptable to change, or less so? Could the resulting forest even survive? A chickadee, though fast in the air, is slow on the ground. Arguing from the bird's perspective, we know that ground speed is not what makes the animal well adapted to its life. Forcing the analogy to its conclusion, we must admit that it is necessary for only some of the animals, even a small percent, to be able to run quickly. So the question becomes, not what are the necessary standards to which we hold every individual, but rather how many individuals must we hold to each of our standards?

It might be useful to see how the new form of our question plays out when applied to a particular piece of mathematics. As a case study, we will take group theory. By group theory, we don't mean the large body of theorems and definitions generally included in a class for mathematics majors, nor do we necessarily mean the facility with proof that we usually expect from those students. Because we are talking about general standards and especially about humanities majors who are generally excluded directly from the standard abstract algebra class, let us be clear about what aspects of group theory we are invoking in this discussion. Here are some manageable standards for such students: (1) the ability to explain what a group is and identify an example as either satisfying or not satisfying the definition of group; (2) a working acquaintance with a variety of examples of groups and their actions on assorted sets; (3) the ability to use some of these examples to accomplish something outside the realm of pure mathematics; and (4) the ability to make simple conjectures about what might be true of groups, to generalize a little, and to investigate those conjectures.

One of the courses developed at Dartmouth, the "Pattern" course, addresses these goals directly. Students investigate the properties of finite and certain infinite groups by constructing associated multiplication tables, noticing patterns in the tables, making conjectures about the properties of these structures, and occasionally being exposed to a mathematical proof of a simple fact, such as "the order of a subgroup divides the order of the group." They do this in the context of design, also creating many of the symmetry groups they are studying in paint, ink, and block print. Through their mathematics, art, and writing, the instructors of the course (one mathematician and one artist) can judge how well the students have met both the mathematical standards listed above and the design standards set for the course.

Notice how these standards, although low for a mathematics major, capture many of the goals that humanists gave for their students' interaction with mathematics. Students did indeed learn to analyze their own art according to abstract mathematical categories. They did integrate aesthetic questions with the academic demand to understand the classification of patterns, and they ended up exploring the cultural context of both art and mathematics far better than students taking the usual courses in either of those majors. Now, the question becomes, What advantage is there for society if a certain number, let us say 1 percent, of humanities majors come away from their education having satisfied these standards?

Taking this question in the most general terms possible, we find that the main effect would be to spread ideas around, in the same way that new genes can be introduced into the gene pool of a species when an isolated population comes into contact with a larger population of the same species. The idea of using the paradigms of both evolution and ecology to frame sociological questions about the propagation of ideas in a culture provides a useful model for considering the culturewide "problem" of "quantitative literacy." This model of information exchange was hinted at in early works by the psychologist Gregory Bateson in the context of individual intelligence (Bateson 1972). Not much later, biologist Richard Dawkins offered a population-wide model, noting that, like genes, ideas interact and combine with one another to produce eventual macroscopic changes in the culture as a whole

(Dawkins 1976). This model for the spread of information is a growing area of research, and the reader is referred to two works by Aaron Lynch (1996a, 1996b) for a taste of this research.

At present, those who study the spread of ideas in a manner analogous to the spread of diseases look at the question of idea propagation as a natural, unassisted phenomenon. They model the propagation of ideas in a (human) culture the same way one might model the growth of bacteria in a (nutrient) culture. Blurring the obvious distinctions between these two situations allows us to apply insights gained from ecological models to the difficult problem of making a particular set of ideas (mathematics, in this instance) more pervasive in a particular culture. What is proposed here is a little social engineering to guide what already happens naturally. Like evolution or a chemical reaction, the spread of ideas is viewed as a process, and purposefully moving ideas around allows this process to happen faster. As with changes in the gene pool, it is hard to predict what measurable differences will result in a population where ideas have recently been mixed up in this probabilistic fashion. But it is possible to offer some hypothetical examples of how advantages might accrue from such a mix.

First let us look at an economic possibility. At present, there are quite a few pieces of software used in computer-assisted design that are based on group-theoretic principles. These are used to generate repeat patterns with one of the seventeen "wallpaper" groups describing their symmetries. It is quite reasonable to expect that within the next five or ten years we will see the development of an ink-jet printer that can be used on fabric. At that point it will be easy for anyone with artistic talent, an understanding of the software, and a small sum of money to start a textile-printing business right out of the garage. Since the economic trend lately has been toward small business and localized production, it would be astonishing if nobody did this. In fact, the country produces such a large number of art majors that if even 1 percent of them had the mathematical and technical expertise to adapt to such an opportunity, the country could quickly come to dominate a brand new aspect of the worldwide textile industry. Nonetheless, such a project would require people who were very good at art and also moderately savvy about group theory; otherwise, such economic gain would be impossible.

Now let us imagine how our 1 percent might benefit the culture at large. The widely lamented split between the more scientifically savvy and the rest of the population is the fault of both sides. Scientists often refuse to speak a language everyone else can understand. Yet nonscientists often avoid the work necessary to understand what the scientists are saying. Our educated humanists are in an excellent position to address both sides of the problem because they not only exemplify the ways in which an average citizen might be scientifically informed but also have the knowledge and background necessary to communicate some of the beauty of science and mathematics to nonscientists. Sol Lewitt's installation art based on algebraic equations; *Arcadia,* a play about chaos and dynamical systems by Tom Stoppard (1993); and an expository volume on art and physics by the physician Leonard Schlain (1991) are all examples of how our culture becomes richer and better informed by mathematically educated humanists. These are all first-order effects of widely dispersed knowledge.

There is also an important second-order effect. One major way that knowledge is passed on in any culture is from parent or teacher to child. For many, one of the early lessons of childhood is to fear mathematics. If mathematical knowledge is dispersed widely throughout the population, no matter what exactly the content is, we can be sure that the opportunities for a child to learn to fear mathematics will decrease. In particular, our 1 percent of humanists who appreciate group theory should feel the way some of the students in the "Pattern" class did. One tells us, "The great thing about courses like Pattern for me [was] that it did give me confidence about math again, but in a different way ... I just learned its place in the universe." Any child taught by this humanities major would be unlikely to fear mathematics and science, particularly if the curriculum were to reinforce such a useful perspective. The long-term value of this second-order effect may be the most desired outcome of all.

Viewed in such a holistic fashion, the needs of the culture need no longer take ideological precedence over the needs of the individual. As in an ecosystem, the individual can be mostly free to pursue his or her own ends, including along the way many parts of mathematics that are relevant to those ends. Our goal should not be the homogeneity of mathematical background for everyone but rather a solid grasp of particular parts of mathematics, with specific content varying widely throughout the population.

QUANTITATIVE LITERACY AND THE GRADES K–12 EDUCATIONAL SYSTEM

An organic view of the role of education in a culture, such as the view offered here, by no means contradicts the need for basic standards set out for the mathematical education of children, such as those the NCTM has worked hard to provide. All branches of mathematics are based on an understanding of number and shape and function and curve, which are as basic to mathematics as reading and writing abilities are to literature. Yet one of the glories of literature is its very diversity, its ability to respond to many varieties of human circumstance and condition. Our education of writers preserves this advantage by insisting on basic competency while simultaneously allowing choice and taste to dictate the many circumstances under which college students write. Mathematics should work the same way at the college level, which at present it most emphatically does not. This is particularly true in our preparation of future mathematics teachers.

Sociobiologists embrace the language of epidemiology to describe their vision of the propagation of ideas throughout a culture, seeing this process as analogous to the spread of disease. In our culture, the primary "vectors" of both the love and hate of mathematics to the next generation are parents or other caregivers and teachers; thus the education of future grades K–12 teachers presents a special quandary for the design of undergraduate courses. It seems quite obvious that a firm grasp of mathematics through the high school level ought to be prerequisite to a career as a teacher of children of any age, yet it is doubtful that we can claim success at even that level. Courses designed to enhance a basic understanding of mathematics with its cultural and aesthetic appreciation assume such an understanding at the outset. These courses could be used effectively to enlarge, but not replace, an already sound teacher preparation program, giving a much needed opportunity for future teachers to "do" mathematics during every year of their undergraduate career in a wide variety of interdisciplinary settings.

QUANTITATIVE LITERACY IN THE TWENTY-FIRST CENTURY

What, then, constitutes the measure of the quantitative literacy of an entire culture? Some part of mathematics should be thought about and read about by the average person with a consistency similar to that individual's consideration of new medical advances, the latest album by their children's favorite rock group, the art exhibit at the local museum. The conversation at a dinner party of twenty should naturally feature some discussion of a mathematical topic. Children should hear about such modern inventions as group theory from their parents, the parents of their friends, their older siblings—who regard the practice of mathematics as pleasurable as reading. The question "Do you like mathematics?" should everywhere be replaced with the question "What is your favorite mathematics?" In the language of the sociobiologists, the entire population should be "infected" with the love of mathematics. In short, the culture should embrace mathematics ancient and modern with enthusiasm and affection. A radical way to put this goal would be to imagine a worst-case scenario: Mathematics should be so deeply embedded in all parts of the culture that if all professional scientists and mathematicians were to disappear from the country overnight, the major areas of mathematical study would still be well represented in the collective understanding of the remaining population.

If we desire such a marvelous outcome, what must we do? It seems from our discussion that a standardized course of study would actually be counterproductive, whereas offering a wide variety of interdisciplinary courses in which students study a specific topic in depth might satisfy society's larger and more pervasive needs. These courses are appealing enough to undergraduates to allow the number of required mathematics courses to increase considerably and still be quite palatable to students. The hard work in a solution of this sort is still left to do, however. College administrations have to sustain us with the resources to make interdisciplinary courses widely available and institute a system of requirements that ensures that students sample them.

Dartmouth's experience shows that a commitment of forty faculty members and their associate deans can be the beginning of a long-term change. The faculty instituted an interdisciplinary requirement for students as well as a quantitative one, either of which can be satisfied by any of the interdisciplinary mathematics and humanities courses. Faculty members work in pairs, a mathematician with a humanist, to develop a course that reflects the interests of both parties. The work demands a creative approach, and both parties usually report learning new things about both their subject material and the teaching of it. The resulting course of study is intellectually satisfying to all concerned and usually somewhat specialized to the demands of the students, instructors, and institution. Students gain greatly from the viewpoints of two very different disciplines, the interaction between faculty, and the range of pedagogical approach to the material. Best of all, a wide range of courses results, allowing students to choose among many different types of mathematics, all presented at the level of novice for those with only a high school background.

All of us who teach mathematics should broaden our understanding so that we can construct the kind of experience for students that makes these sorts of courses work. We have to learn to think about mathematical things the way an artist, a musician, a philosopher, or a writer might, so that all our students can learn to think like mathematicians.

References

Bateson, Gregory. *Steps to an Ecology of Mind*. San Francisco: Chandler Publishing Co., 1972.

Dawkins, Richard. *The Selfish Gene*. New York: Oxford University Press, 1976.

Lynch, Aaron. *Thought Contagion*. New York: Basic Books, 1996a.

———. Chapter 1 of *Thought Contagion*. Available online: www.mcs.net/~aaron/tc1.html, 1996b.

Schlain, Leonard. *Art and Physics: Parallel Visions in Space, Time, and Light*. New York: William Morrow & Co., 1991.

Stoppard, Tom. *Arcadia*. Boston: Faber & Faber, 1993.

4

The Standards Movement in Mathematics Education
Reflections and Hopes

Joan Ferrini-Mundy

WITH the release of the *Curriculum and Evaluation Standards for School Mathematics* in 1989, the National Council of Teachers of Mathematics (NCTM) led the way into new territory for content-based professional organizations. The mathematics education standards odyssey, now more than a decade long, has included some highlights that were anticipated and consequences that were unanticipated and has given rise to an evolving and challenging set of issues that extend far beyond the confines of grades K–12 mathematics classrooms. The following discussion will address elements of the history of mathematics education standards, summarize some of what is known about the effects and influence of mathematics education standards, chronicle the process of revising standards, and address some of the features of *Principles and Standards for School Mathematics: Discussion Draft* (NCTM 1998). Within each section, issues that are relevant to the yearbook's theme, "Learning Mathematics for a New Century," will be highlighted.

BACKDROP AND HISTORICAL HIGHLIGHTS: MATHEMATICS EDUCATION STANDARDS

NCTM leaders recall finding themselves in quite unfamiliar territory in the spring of 1989, standing before floodlights and the national press in Washington, D.C., at the release of the *Curriculum and Evaluation Standards for School Mathematics*. The *Standards* was the result of both national pressure to reform mathematics and science education and a view within the NCTM

This paper was prepared before the final version of *Principles and Standards for School Mathematics* was available in 2000. The material in this paper is based on *Principles and Standards for School Mathematics: Discussion Draft* (NCTM 1998).

leadership and committee structure of a need for a statement by NCTM that would "draw together the best knowledge and experience of the profession into a national statement of standards and ... devise effective mechanisms for the implementation and continual review of those standards" (Crosswhite as quoted in McLeod et al. 1996, p. 34). A document of this type was new to the field and ultimately generated reaction that reached further beyond the community than the original writers could possibly have imagined.

The idea of comprehensive mathematics education standards focusing on curriculum, instruction, and evaluation originated as a result of trends in the late 1970s and early 1980s toward "back to basics" (Crosswhite 1990), public concern about declining test scores (NCTM 1980), and the warning in *A Nation at Risk* (NCEE 1983) about a rising tide of mediocrity and the need for high standards in all subjects (McLeod et al. 1996). The report of the National Advisory Committee on Mathematical Education (NACOME 1975) observed the emergence of false dichotomies in mathematics education: "false choices between the old and the new in mathematics, skills and concepts, the concrete and the abstract" (p. 136). In this climate, NCTM leaders began to consider playing a stronger role in defining the nature and future of school mathematics. In 1980, *An Agenda for Action* (NCTM 1980) made recommendations for school mathematics and conveyed that NCTM wanted to "provide direction for the field, to assert its authority and share its expertise with a higher level of intensity than had been its custom" (McLeod et al. 1996, p. 24).

A series of events in the early 1980s seemed to converge to generate first the concept and then the development of standards. There were reductions in federal funding for science and mathematics education, and the National Science Board issued a report calling for consensus about new objectives for the precollege curriculum and for "guidelines for new curricula" (National Science Board 1983, p. 46). At a meeting of the Conference Board of the Mathematical Sciences in 1983, Joe Crosswhite proposed the idea that "there should be a set of standards for school mathematics prepared by NCTM and that he wanted to get the whole group behind him to support this" (Steen as quoted in McLeod et al. 1996, p. 28). A subsequent meeting at the University of Wisconsin—Madison produced a similar recommendation for the development of guidelines for the mathematics curriculum. NCTM's Instructional Issues Advisory Committee put forward the initial plan for standards development, suggesting the production of a document during 1985–86. In fact, the process began in 1986 with the appointment of a Commission on Standards for School Mathematics and in the summer of 1987 with the first full meeting of the writers. A draft was released in October of 1987, and following national review and discussion, it was revised in the summer of 1988. The *Curriculum and Evaluation Standards for School Mathematics* was released in March 1989. (A history of the *Standards,* including the *Professional Standards for Teaching Mathematics,* is available in McLeod et al. [1996].)

WHAT HAPPENED: EFFECTS OF THE *STANDARDS*

The *Curriculum and Evaluation Standards* was designed to speak to those very close to decisions about mathematics curriculum—teachers, supervisors, and developers of instructional materials and curricula. Ultimately, the trio of *Standards* documents (NCTM 1989, 1991, 1995) took on a "life of its own" and had influence, sometimes unexpected, in several arenas. Some highlights of this period should be mentioned, since they helped shape the foundation on which the *Standards* documents were eventually revised. The National Science Foundation (NSF), concurrent with the launching of the standards movement, was initiating the systemic reforms in states, cities, rural areas, and other local regions. In the early Systemic Initiative requests for proposals, the NCTM *Standards* documents were cited as the mathematics curriculum framework that was to be promoted in systemic reforms. During the same period, a major NSF curriculum development program in mathematics was instituted, with the *Standards* again serving as the framework. During this period, the *Standards* were received as a major resource that would guide teacher enhancement efforts, curriculum development, and classroom practice.

Interestingly, the matter of how the *Standards* might influence the classroom practice of teachers directly, which was more of the original intention, was eclipsed to some extent as the policy community took such a keen interest in the documents. A byproduct of their adoption by their policy community is the finding that their influence on classroom practice was, by and large, mediated through the translation of standards into other forms. Teacher education programs attempted to enact the ideas of standards and to help teachers understand and use them; curriculum materials—ranging from the NCTM Addenda series to the NSF-supported K–12 curriculum projects to the materials produced by commercial publishers—all attempted to provide enactments of the *Standards* ideas and thereby influence classroom practice. State curriculum frameworks (Council of Chief State School Officers 1997) were developed or revised to reflect *Standards* ideas. The effect of the *Standards,* although intended for classroom practice, may in fact have been much more indirect and far-reaching.

So what features from the original *Standards* documents seemed to make the most impact? Unfortunately, there has been no systematic accumulation of research data to help answer this question. Some studies do suggest that pedagogical features of the *Standards* (emphasis on cooperative groups, emphasis on writing in the mathematics classroom, emphasis on discussion and discourse) were more readily taken up by teachers than some of the mathematics-content features (Ferrini-Mundy and Johnson 1997).

As the Crosswhite quotation indicates, from the very earliest days of standards discussions, NCTM leaders saw this as an ongoing, evolving process of continual revision. Bybee (1997) has characterized the standards cycle as shown in figure 4.1. With this perspective, standards can be thought of as a tool for the field to use in various ways.

The concept of standards has proved to have several interesting features. Standards *contribute language and concepts* into the field for discussion and reference. There are a number of examples from the original mathematics education standards that could be mentioned; for example, a focus on discourse in mathematics classrooms has given teachers and educators a way of talking about communication goals for mathematics education. In addition, the choice of standards *can draw attention to particular concepts and ideas*. A good example of this is the inclusion in the *Curriculum and Evaluation Standards* of discrete mathematics as a content area needing emphasis,

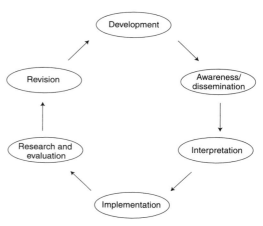

Fig. 4.1. Standards cycle

and of connections as a process standard. In fact, the notion of "process standards" was introduced into the common discussion following the release of the *Standards*. Standards also serve a purpose of *bringing into focus important areas of disagreement*, such as the role of paper-and-pencil symbol manipulation or the place of calculators in learning about number.

Another unanticipated outcome was the debate and critique that would follow the *Standards* years past their initial publication. As curriculum materials purporting to be consistent with the *Standards* became widely available and as teachers became familiar enough with the document to claim they were using it, the field started to react, often with rather negative criticism (e.g., Wu 1997). Responding to this issue of interpretation, Bass (1998) stated: "One of the hazards of standards, especially at the national level, is that the general language that must be employed admits widely varying interpretations, sometimes very much at variance with what the framers of the standards intended" (p. 4).

Even though some people have claimed the original 1989 *Standards* document was too vague, the fact is that these standards were specific enough for people to find areas to critique. Many perceived the *Standards* as overemphasizing such notions as "real world" applications or teaching mathematics through problem solving. The documents were viewed by some as not emphasizing accuracy and mathematical knowledge.

STANDARDS 2000: REVISING THE NCTM *STANDARDS* DOCUMENTS

Since the mid-1980s there has been an NCTM committee with responsibility for thinking about standards-related issues and the organization's plans for promoting standards ideas. This committee, then called the Commission on the Future of the Standards, produced in 1996 a proposal to the NCTM Board of Directors for what was called "The Synthesis and Update Project." A group would be appointed to establish standards that—

- build on the foundation of the original *Standards* documents;
- integrate the classroom-related portions of the *Curriculum and Evaluation Standards for School Mathematics,* the *Professional Standards for Teaching Mathematics,* and the *Assessment Standards for School Mathematics;*
- are based on four grade bands: pre-K–2, 3–5, 6–8, and 9–12.

The ensuing effort soon became known as the Standards 2000 project and included the commission, a Writing Group, and an Electronic Format Group. The addition of an electronic format version of the standards is a particularly exciting element of this activity.

The Writing Group was appointed in 1996–97, with very careful attention to the inclusion of classroom teachers, teacher educators, mathematics education researchers, mathematicians, developers of instructional materials, and state and local mathematics education policymakers. Each of the four grade-band teams included a classroom teacher and a mathematician, along with representatives of the other groups. The Standards 2000 Writing Group first met informally at the NCTM Annual Meeting in 1997 and then for a working session in Pleasant Hill, California, in the summer of 1997. As background, the writers had access to Association Review Group (ARG) responses, curriculum materials, state and international mathematics frameworks, research articles, and other resources that could be useful.

The original writers of the *Curriculum and Evaluation Standards* faced the daunting challenge of inventing meanings and images for standards. The group vacillated between the idea of standards as criteria to judge quality and the idea of standards as a vision for what might be possible (McLeod et al. 1996). The Standards 2000 writers faced a different challenge—that of building on the existing documents while at the same time clarifying the messages, responding to what the field had learned in ten years, incorporating input from a range of sources, meeting the needs of the field as they stood after ten years with the *Standards,* and providing enough inspiration to help move mathematics education forward in the twenty-first century.

The Commission on the Future of the Standards was responsible for orchestrating an elaborate set of mechanisms for input to the development of Standards 2000. A major source for input was the ARGs. Each of the member societies of the Conference Board of the Mathematical Sciences was invited to form a committee that would supply input to the Writing Group throughout the development process for the standards. Over a two-year period, fourteen ARGs were formed and participated in the process. Periodically, questions from the Writing Group were forwarded to the ARGs for their input. For example, in the second round of questions, the ARG members were asked to respond to the following questions:

- What mathematical reasoning skills should be emphasized across the grades?
- How should the standards address mathematical proof?

ARG responses were synthesized and furnished to the writers. The feedback from these groups influenced the messages and structure of *Principles and Standards*. In turn, there is some evidence that the process of considering the NCTM questions has generated substantial discussion and exchange of ideas about grades K–12 mathematics education among the ARG communities. In addition, special white papers were commissioned summarizing research about students' learning in several areas of mathematics. The electronic format and the availability of the works of ARGs and white papers provides a rich backdrop for the standards and make clear the "national" community (not just the NCTM) that was involved in their development. A conference on the Foundations of School Mathematics offered additional research and theoretical foundation. The white papers and conference papers together constitute *A Research Companion to the NCTM "Standards"* (Kilpatrick, Martin, and Schifter forthcoming). Technical advisory resource people were commissioned to provide very specific guidance in areas including equity and technology. In general, the calls for input and advice during the initial development process included a wide range of stakeholders in mathematics education, especially mathematicians.

With this substantial input, and after two summers of extensive and difficult discussion and writing, *Principles and Standards for School Mathematics: Discussion Draft* (NCTM 1998) was released in October 1998. The electronic version, featuring additional dynamic examples, was released in January 1999.

The input that reached the Writing Group conveyed a wide range of opinions, almost always with equally articulate cases made for both sides of a particular issue. Resolving the tensions in each issue area became a challenge for the Writing Group. The sections that follow illustrate several of the challenges that emerged.

Make the Document More Detailed vs. Preserve the Same Level of Generality That Was in the 1989 *Standards*

In the years following the release of the *Curriculum and Evaluation Standards*, states and districts used the document as a guide for the development of state frameworks and curriculum guides. These documents often provided grade-by-grade, topic-by-topic details about what students should know and be able to do. Such lists generated great debate and controversy in some states, and considerable pressure was put on the Standards 2000 Writing Group to "settle" the question of grade-by-grade expectations by providing explicit detail. Kirst and Bird (1997, p. 17) have noted: "If you approve standards that are too general, or do not contain pedagogy, you will be criticized that there is insufficient instructional guidance for teachers, and the content gaps will be filled by tests or assessments. If you do approve pedagogy or detailed standards, you will be criticized because standards are too long, complex, and overly control local practice."

The solution that the Writing Group reached can be viewed as something of a middle road between the broad stance of the *Curriculum and Evaluation Standards* and the state-level detail. The draft *Principles and Standards* presents ten standards that span the grades and then offers lists of specific topics and instructional emphases within each of the grade-band sections. It is the view of the Writing Group that specific decisions about sequencing and grade-level goals must generally be made locally with attention to local contexts and goals.

Emphasize the Learning of Mathematical Content vs. Emphasize Process and the Idea of Valuing Mathematics

A major part of the controversy and debate about mathematics education in the late 1990s involved accusations that the NCTM was promoting "fuzzy math" (Cheney 1997) through its *Standards* and the related publications. Critics were concerned that a stronger emphasis on the learning of mathematical content was needed and that the following "New Goals for Students" (NCTM 1989, pp. 5–6) were not focused enough on the actual learning of mathematics:

1. Learning to value mathematics
2. Becoming confident in one's own ability
3. Becoming a mathematical problem solver
4. Learning to communicate mathematically
5. Learning to reason mathematically

Responses to this concern in the draft *Principles and Standards* are actually fairly subtle. One decision made by the Writing Group was to place the "content" standards first. These draft standards were divided into five content categories: Number and Operation; Patterns, Functions, and Algebra; Geometry and Spatial Sense; Measurement; and Data Analysis, Statistics, and Probability. The content standards would be followed by the "process" standards, thereby making it possible to mention content areas specifically in the process standards discussions. The process standards are also divided into five categories: Problem Solving, Reasoning and Proof, Communication, Connections, and Representation (NCTM 1998, pp.. 45, 46). In addition, there was some attempt to frame the process standards more directly in relation to mathematical learning outcomes. Note the following comparison of the language from the 1989 *Standards* and the 1998 *Principles and Standards* draft in figure 4.2.

Curriculum and Evaluation Standards for School Mathematics (NCTM 1989)	Principles and Standards for School Mathematics: Discussion Draft (NCTM 1998)
PROBLEM SOLVING (9–12)	PROBLEM SOLVING (pre-K–12)
Use, with increasing confidence, problem-solving approaches to investigate and understand mathematical content. (p. 137)	Build new mathematical knowledge through their work with problems. (p. 49)
REASONING (K–4)	REASONING (pre-K–12)
Believe that mathematics makes sense. (p. 29)	Recognize reasoning and proof as essential and powerful parts of mathematics. (p. 49)
COMMUNICATION (5–8)	COMMUNICATION (pre-K–12)
Appreciate the value of mathematical notation and its role in the development of mathematical ideas. (p. 78)	Use the language of mathematics as a precise means of mathematical expression. (p. 49)

Fig. 4.2

These small but intentional changes are meant to suggest a stronger emphasis on mathematics learning. Some of the reaction to the *Principles and Standards* draft has indicated on the one hand a sense that process has been de-emphasized by the placement of the standards. On the other hand, reaction from some mathematicians seems to indicate that they believe content has been more effectively emphasized. The main point is that all four of the original process standards have been preserved, another ("Representation") has been added, and they span the grades.

Address the NCTM *Standards* Documents to Teachers vs. Address the NCTM *Standards* Documents to Policymakers

The 1989 *Standards* document was written to "each of you concerned with the teaching and learning of mathematics" (p. 12), and it was distributed to the entire NCTM membership. The way in which policymakers at all levels (federal agency staff, state and district supervisors and administrators, curriculum developers, and so on) chose to use the document as a basis for their work had not been fully anticipated. As a result, it became clear over the ten-year period before *Principles and Standards* that in addition to serving as guides and resources for teachers, standards serve as policy-making tools. Thus the question of audience became significant for the writers of *Principles and Standards*. For the draft version, our intentions were quite mixed. An oversimplification of the situation is that the first three chapters of the draft are aimed more at policymakers. In fact, the inclusion of the new chapter, "Guiding Principles for School Mathematics Instructional Programs" (NCTM 1998, pp. 21–43), was intended to speak directly to those at all levels who formulate policies and have responsibility for school mathematics instructional programs. (The intended audience for the final document is teachers and all who make decisions about mathematics education.)

The grade-band chapters that follow the introductory material are written more explicitly with teachers in mind and include classroom episodes, instances of students' work, and ideas about teaching the proposed mathematics.

Promote Evolutionary Change vs. Provide a Bold Vision for the Future

Developing a document that is "visionary" can mean providing unfamiliar and revolutionary ideas for people to rally behind. In many senses the 1989 *Curriculum and Evaluation Standards* achieved this type of vision. As a result, a number of areas—skills, concepts, practice, technology, students' invention, and so on—were given "all or none" interpretations by some. And so, with ten years of experience in watching how these interpretations arose, the Standards 2000 Writing Group faced the interesting challenge of determining how to address some of these problematic interpretations in ways that indicated how more-balanced approaches might be useful and at the same time preserving and reinforcing the important contributions and new directions offered by the original *Standards*. Thus, *Principles and Standards: Discussion Draft* includes statements that are perceived by some as being too equivocal and not strongly visionary. This was a continuing dilemma for the writers as they prepared the final draft of these standards, although we were strongly influenced by advice to "be bolder."

Include Examples to Show "What It Should Look Like" vs. Include Examples to Promote Thought and Deeper Understanding

The Writing Group has faced an interesting challenge in choosing and developing examples for both the written version and the electronic version. In many ways, standards documents can be envisioned as resources for teacher reflection, discussion, and learning. With this purpose in mind, the examples that might be chosen would go beyond illustrations of "standards-based practice" and could incorporate discussion about the challenges and unexpected dilemmas that arise in the teaching and learning processes. The electronic format particularly allows for examples that are extended and can be used and viewed in a range of ways. The tension here is that if standards are supposed to present a clear vision of what is intended in school mathematics, then this clarity is potentially blurred by indicating complexity and different possibilities within examples.

THE DRAFT "PRINCIPLES AND STANDARDS": WHAT IS NEW?

The Writing Group was charged with "building on the foundations" of the original *Standards* documents while at the same time taking into careful account the substantial input coming from many parts of the mathematics and education communities. During the project, the results of the Third International Mathematics and Science Study (TIMSS) were released, and the characterization of the mathematics curriculum in the United States as being a "mile wide and an inch deep" (Schmidt, McKnight, and Raizen 1997) was repeatedly discussed. Schmidt and his colleagues found that our mathematics curricula, in the form of textbooks and state frameworks, included far more topics each year than the corresponding curricula in other countries.

The Writing Group attempted to address this concern in two ways. First, it proposed a set of ten standards that would span the grades pre-K–12 spectrum. This meant that the group needed to reach agreement on the language of standards, and in particular the focus areas within standards, so that each of the four grade-band groups would be able to interpret the focus areas in a way appropriate to their grade band. The Writing Group also agreed that for some grade bands, some focus areas within a particular standard might not be addressed at that grade band. In addition, *Principles and Standards* attempts in some areas to be quite specific about when students reach "closure" with certain topics and move on. This is discussed in the draft Mathematics Curriculum Principle: "The problems posed and the concepts examined within any mathematics area should grow more sophisticated each year of the curriculum until a suitable level of understanding or proficiency is reached. At that point, although formal instruction in the specific content area will end, students should continue to use the understandings and skills they have acquired, thereby maintaining and strengthening both" (NCTM 1998, p. 30). Throughout the draft there are efforts to describe when students should reach levels of proficiency in certain areas.

Perhaps more significant than the idea of "closing off" the study of particular topics is the notion that all the mathematical strands proposed in *Principles and Standards* should, in some way, be developed over the years of schooling. This perspective embodies the notion of learning through building on previous knowledge. In the overviews in chapter 3, *Principles and Standards* attempts to present the developmental trajectory of each of the ten standards areas across the grades. The trajectories described there are then supported in the grade-band descriptions that follow.

This notion of developing deeper knowledge that builds on previous knowledge across the grade span is somewhat more familiar for the content areas, especially number and, increasingly, algebra. In fact, in algebra a number of efforts have been made to describe the development of algebra across the grades (National Research Council 1998). Describing the development within the process areas across the grades proved to be an especially challenging enterprise.

To illustrate how *Principles and Standards* attempts to emphasize the notion of building across the grades, consider the standard in figure 4.3 (NCTM 1998).

Standard 2: PATTERNS, FUNCTIONS, AND ALGEBRA

Mathematics instructional programs should include attention to patterns, functions, symbols, and models so that all students—
- understand various types of patterns and functional relationships;
- use symbolic forms to represent and analyze mathematical situations and structures;
- use mathematical models and analyze change in both real and abstract contexts.

Fig. 4.3

In each grade-band chapter there is a discussion of each of these three focus areas. Tracing the examples used for the second focus area, "Use symbolic forms to represent and analyze mathematical situations and structures," indicates how teachers might observe and plan for growth in understanding across the grades. In grades pre-K–2, a child's invented notation for the number ten and a half is shown in figure 4.4.

This notion of building on children's strategies is illustrated again in the grades 3–5 section, with more sophistication and with movement closer to conventional mathematical ideas through the following example: "Students might multiply 8 and 14 by using the distributive property; 8×10 plus 8×4" (NCTM 1998, p. 167). In the middle grades, this line of thinking becomes more advanced, and the following problem situation is offered as an example of how teachers might introduce algebraic expressions as models

Fig. 4.4. A child's invented notation (NCTM 1998, p. 119)

of quantities in contextual situations: "Suppose three friends attend a ball game and each purchases a hot dog and a soda. The total cost could be denoted 3 $(H + S)$, where H is the cost of a hot dog and S is the cost of a soda. But it is also possible to think separately about the cost of the hot dogs $(3H)$ and the cost of the sodas $(3S)$. In this case, the total would be denoted $3H + 3S$" (NCTM 1998, p. 225).

By grades 9–12, students' abilities to represent situations algebraically should become quite sophisticated. (See NCTM 1998, pp. 286–88.)

REACTION, SYNTHESIS OF REACTION, AND FINAL REVISION

The commission and the Writing Group were deeply involved in synthesizing and responding to the substantial feedback and reaction that was gathered in the period between October 1998 and May 1999. ARG reviews, commissioned reviews, responses provided by groups of teachers, and individual responses directed to NCTM numbered in the hundreds. These responses were coded electronically using methods from qualitative research data analysis, and a variety of reports on the major issues emerging through the feedback process were produced for the writers and others. This work was conceptualized and carried out by Gary Martin, NCTM director of research, and Mary Lindquist, chair of the commission, with support from the commission and others. The National Research Council conducted a review of the revision process that examined the adequacy of the Writing Group's response to the input. In the summer of 1999, the writers produced the final manuscript of *Principles and Standards for School Mathematics.*

CONCLUSION

The effect of standards on grades K–12 mathematics education in the past decade has been significant in focusing conversation and effort within mathematics education in a common direction. The debates and critique that have been precipitated by standards work can be taken as a positive indication of the extent to which mathematicians, educators, parents, and school personnel care about shaping the most effective mathematics instructional experiences possible for our students. With the release of *Principles and Standards for School Mathematics* in April 2000 along with this yearbook, we hope to provide continuing resources for the improvement of mathematics teaching and learning for all students. If *Principles and Standards,* together with its electronic version, can serve to deepen the ongoing discussions and efforts to provide the best mathematics education possible for students at the beginning of the new century and if ten years hence a new NCTM group is charged with developing a next document that "builds on the foundations of *Principles and Standards,*" then standards will have become a central resource and focus for achieving quality in mathematics education.

REFERENCES

Bass, Hyman. "The Education Debates: A Discussion Paper for CSMEE." Paper prepared for the National Research Council, Center for Science, Mathematics, and Engineering Education. Washington, D.C.: National Research Council, 1998.

Bybee, Roger W. "A Strategy for Standards-Based Reform of Science and Mathematics Education." Unpublished manuscript. Washington, D.C.: National Research Council, 1997.

Cheney, Lynne. "Once Again, Basic Skills Fall Prey to a Fad." *New York Times,* 11 August 1997, sec. A, p. 15.

Crosswhite, F. Joseph. "National Standards: A New Dimension in Professional Leadership." *School Science and Mathematics* 90 (October 1990): 454–66.

Council of Chief State School Officers. *Mathematics and Science Content Standards and Curriculum Frameworks: State Progress on Development and Implementation—1997.* Washington, D.C.: Council of Chief State School Officers, 1997.

Ferrini-Mundy, Joan, and Loren Johnson. "Highlights and Implications." In *The Recognizing and Recording Reform in Mathematics Education Project: Insights, Issues, and Implications,* edited by Joan Ferrini-Mundy and Thomas Schram, pp. 111–28, *Journal for Research in Mathematics Education* Monograph No. 8. Reston, Va.: National Council of Teachers of Mathematics, 1997.

Kilpatrick, Jeremy, W. Gary Martin, and Deborah Schifter, eds. *A Research Companion to the NCTM "Standards."* Reston, Va.: National Council of Teachers of Mathematics, forthcoming.

McLeod, Douglas B., Robert E. Stake, Bonnie P. Schappelle, Melissa Mellissinos, and Mark J. Gierl. "Setting the Standards: NCTM's Role in the Reform of Mathematics Education." In *Bold Ventures,* Vol. 3, *Case Studies of U.S. Innovations in Mathematics Education,* edited by Senta A. Raizen and Edward D. Britton, pp. 13–132. Boston: Kluwer Academic Publishers, 1996.

National Advisory Committee on Mathematical Education. *Overview and Analysis of School Mathematics, Grades* K–12. Washington, D.C.: Conference Board of the Mathematical Sciences, 1975.

National Council on Excellence in Education. *A Nation at Risk: The Imperative for Educational Reform.* Washington, D.C.: U.S. Government Printing Office, 1983.

National Council of Teachers of Mathematics. *An Agenda for Action: Recommendations for School Mathematics of the 1980s.* Reston, Va.: National Council of Teachers of Mathematics, 1980.

———. *Assessment Standards for School Mathematics.* Reston, Va.: National Council of Teachers of Mathematics, 1995.

———. *Curriculum and Evaluation Standards for School Mathematics.* Reston, Va.: National Council of Teachers of Mathematics, 1989.

———. *Principles and Standards for School Mathematics: Discussion Draft.* Reston, Va.: National Council of Teachers of Mathematics, 1998.

———. *Professional Standards for Teaching Mathematics.* Reston, Va.: National Council of Teachers of Mathematics, 1991.

National Research Council. *High School Mathematics at Work: Essays and Examples for the Education of All Students.* Washington, D.C.: National Academy Press, 1998.

National Science Board. *Educating Americans for the Twenty-first Century.* Washington, D.C.: U.S. Government Printing Office, 1983.

Schmidt, William H., Curtis C. McKnight, and Senta A. Raizen. *A Splintered Vision: An Investigation of U.S. Science and Mathematics Education.* Boston: Kluwer Academic Publishers, 1997. (The *Executive Summary* of this publication is available on the Web at ustimss.msu.edu/splintrd.htm.)

Wu, Hung-Hsi. "The Mathematics Education Reform: Why You Should Be Concerned and What You Can Do." *American Mathematical Monthly* 104, no. 10 (1997): 946–54.

5

Calculators in Mathematics Teaching and Learning
Past, Present, and Future

Bert K. Waits

Franklin Demana

THE 1986 National Assessment of Educational Progress (NAEP) revealed that 21 percent of the middle school students in the study and 26 percent of the senior high school students in the study attended schools that had calculators available in the mathematics classroom (Dossey et al. 1988, p. 79). In the 1992 NAEP study, these percents had risen to 81 percent and 92 percent, respectively (Dossey and Mullis 1997, p. 26). In 1990, NAEP results showed that 33 percent of eighth graders in public schools were allowed to use calculators for mathematics tests. By 1996, this percent had risen to 70 percent of all eighth graders (Shaughnessy, Nelson, and Norris 1998, pp. 93–95, 191), and nearly 60 percent of eighth graders were using calculators in their mathematics classes on a daily basis! Seldom in the history of mathematics education has such a rapid change been made with such significant consequences. In this article we would like to step out of the rushing stream of this technology-induced change, pause to reflect on our experiences of the past twenty-five years, and describe what we see on the horizon for calculators in mathematics education in the twenty-first century.

We begin with a brief history of calculator development and give some lessons learned about using calculators in our work dating back to the 1970s. A statement of our position on the appropriate use of calculator technology is presented along with a discussion of some of the controversial issues that have arisen as a consequence of the use of calculators. We describe the importance of a balanced approach to the teaching and learning of mathematics that uses both technology and paper-and-pencil techniques. Some research evidence about calculator use is given. Finally, we discuss the status of the use of calculators in the teaching and learning of mathematics in the rest of the world, describe some recent advances in calculator technology, and speculate about the future impact of these advances.

A Brief History

According to Ball (1997), handheld electronic calculators were first introduced to the world by Canon, Inc., in this 14 April 1970 press release from Japan:

> Canon Inc., in close collaboration with Texas Instruments Inc. of the United States, has successfully developed the world's first "pocketable" battery-driven electronic print-out calculator with full large-scale integrated circuitry.

In 1972, Hewlett-Packard introduced the remarkable HP-35, the first "scientific" calculator that evaluated the values of transcendental functions such as log 3, sin 3, and so on. The last slide rule was manufactured in the United States in 1975! In 1986, Casio of Japan introduced the first so-called graphing calculator with powerful built-in, computer-like graphing software. In 1996 Texas Instruments introduced the TI-92, the first calculator that contained an easy-to-use computer algebra system (CAS) *and* a version of Cabri computer interactive geometry (Waits and Demana 1996). Recently, both Texas Instruments and Casio introduced flash ROM calculators, which have many positive implications for the future. Flash technology will enable many kinds of useful computer programs to run on calculators as well as provide easy calculator software upgrades electronically. This feature alone could revolutionize the applicability of calculators in the twenty-first century.

What We Learned about Using Calculators in Mathematics Teaching

After twenty-five years of using handheld calculators, we have learned some fundamental principles about the use of calculators in the teaching and learning of mathematics (Waits and Demana 1994).

We have learned important lessons about change. Arguably the most important thing we learned has to do with desktop computers and why they had very little direct impact on the teaching and learning of school mathematics. We tried using desktop computers in the 1980s in our early projects in which we used our own computer graphing software (i.e., Master Grapher [Waits and Demana 1987]) to enhance the understanding of precalculus and calculus. Whereas teachers became very excited about the possibilities, most students in most schools, we discovered, had very limited, if any, access either to desktop computers or to mathematics computer software in their mathematics classrooms. We found that our ideas were not being used. Our work had very little impact on classroom instruction.

When graphing calculators were introduced, we saw an obvious opportunity because they were very inexpensive, handheld (fit in a shirt pocket), and very computer-like. We immediately began to instruct the teachers in our projects in the use of graphing calculators. The rest is history, as they say. Graphing calculators soon became very popular in many countries, including the United States. The reasons were obvious: every classroom could be turned into a computer lab, and every student could own his or her own inexpensive personal computer with built-in mathematics software (Demana and Waits 1992). We note that the same dynamics are true today for CAS. For example, graphing calculators now have computer algebra systems almost as powerful as personal computer–based software like Mathematica or Maple.

What we have learned does not imply that desktop computers are not important in education! All the software and functionality available today on advanced calculators first appeared on desktop computers. In many ways the calculator borrows what has been proved effective on the computer and makes it accessible to many more students. The lesson we learned is that *change can occur if we put the potential for change in the hands of everyone.* The handheld calculator does precisely this for the mathematics teacher and student. The result is clearly demonstrated by taking a cursory look at the data presented in our opening paragraph.

The second most important thing we learned about change has to do with effective professional development in the use of technology. The adoption and use of technology requires additional teacher in-service training that addresses conceptual and pedagogical issues. However, our early professional development methods consisted largely of demonstrating the use of technology to the large groups of teachers we brought to Ohio State University. This form of teacher in-service training was not good enough. We cannot expect teachers to make fundamental change in their teaching without adequate, ongoing support. Teachers consistently request intensive start-up assistance and regular follow-up activities. But the greater lesson we learned is that teaching in the grades K–12 arena is a profession whose constraints are so complex and abundant that teaching practice is very difficult to change from the outside. Change has to come from within the teaching profession and be supported both from within and from without.

Changing practice is full of local issues that must be dealt with by teachers at that level. Our early top-down, one-dimensional model of professional development simply had no hope of producing the change we wanted to extend to all schools and all mathematics teachers. The best thing we ever did was to turn the professional development activities of our projects over to practicing teachers who had succeeded in embedding the appropriate use of calculators into their own practices.

We have learned that on a large scale, *it takes practiced teachers to change the practice of teachers.* The Teachers Teaching with Technology (T³) program that we founded in 1985–86 is an example of such a professional development program. The T³ program has consistently tried to embody the tried-and-true principles of effective professional development, many of which we learned the hard way in our projects. (For an excellent analysis of effective professional development experiences, see Loucks-Horsley et al. [1998].)

T³ offers intensive teacher education institutes, and the regional and annual T³ meetings afford teachers opportunities to obtain ongoing professional development. Practicing teachers in the T³ institutes model appropriate calculator use in teaching specific mathematics and science topics. The T³ organization also supports its institutes by providing an extensive Web page filled with resources for teachers using calculators in their classrooms as well as areas for discussion, where teachers can get help and share ideas (www.t3ww.org/t3).

We have learned that calculators cause changes in the mathematics that we teach. These changes often can be very dramatic, as we learned from our personal teaching experiences before and after calculators were available. For example, some paper-and-pencil applications have simply become obsolete, as illustrated by the following examples:

- Complicated arithmetic computation: Compare computing
$$\frac{1789}{1.0725}$$
by paper-and-pencil long division with computing the quotient using a simple four-function calculator.
- Interpolation using transcendental-function tables: Compare computing $1250(1.04125)^{12}$ using precalculator, paper-and-pencil logarithmic interpolation with computing the product using an inexpensive scientific calculator.
- Accurate graphing of complicated functions: Compare graphing
$$f(x) = \frac{x^3 - 17x + 7}{x^2 + 1}$$
using traditional paper-and-pencil calculus methods with graphing the function using a graphing calculator (see fig. 5.1). In the past the traditional methods for "graphing" included finding the derivative, $f'(x)$, and solving the equation $f'(x) = 0$ by paper-and-pencil methods only. And only contrived graphing problems with accessible paper-and-pencil solutions would be given. Now students can graph far more functions more accurately than before. Also students can

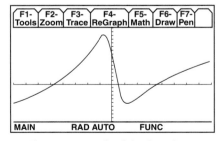

Fig. 5.1. A graph of the function
$$f(x) = \frac{x^3 - 17x + 7}{x^2 + 1}$$

use traditional calculus methods to *confirm analytically* that the graph they see is accurate.

- Complicated integrations: Compare computing the value of the definite integral

$$\int_0^{\frac{\pi}{3}} x^2 \sin(x)\,dx$$

using paper-and-pencil methods with computing the value using a state-of-the-art CAS calculator (see fig. 5.2).

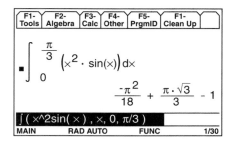

Fig. 5.2. Using the TI-89 to compute the *exact* value of an integral

- Solving complicated equations: Compare finding the real and complex solutions to the simple cubic polynomial equation $3x^3 + 2x^2 - 7x + 9 = 0$ by paper-and-pencil methods (go ahead and try!) with using a calculator-based graphical or numerical method.

We have learned that before calculators, we often asked students to solve only contrived problems. The students, therefore, learned methods that, at least in their minds, were really applicable only in contrived contexts. Calculators allow students to apply more-general types of solution processes even to problems that have no exact solution or to problems that cannot be solved by traditional paper-and-pencil methods alone, as illustrated in the next examples.

- Problems not solvable by paper-and-pencil methods taught in schools: Compare computing the definite integral

$$\int_1^2 \frac{\sin(x)}{x}\,dx$$

with paper and pencil to computing the solution using a calculator with integration functionality. Can you find the indefinite integral as an elementary function (Demana and Waits 1994)?

- Methods too cumbersome before calculators: The parametric graphing utility on most graphing calculators makes possible mathematical modeling and simulation to illustrate and solve problems that were impossible with paper and pencil alone.

Clearly we can solve many more problems using calculators! They facilitate problem solving. In general, when it comes to mathematics, we have learned that the visionary statements made by the eminent mathematician Henry Pollak are very true (Pollak 1986, pp. 347–48). To paraphrase, he said that because of technology—

- some mathematics becomes less important (like many paper-and-pencil arithmetic and symbol-manipulation techniques);
- some mathematics becomes more important (like discrete mathematics, data analysis, parametric representations, and nonlinear mathematics);
- some new mathematics becomes possible (like fractal geometry).

We have learned that calculators cause changes in the way we teach and in the way students learn. Before computers and calculators, it was necessary for students to spend time mastering and becoming proficient in the use of paper-and-pencil computational and manipulative techniques. Today much of this time can be spent on developing deeper conceptual understanding and valuable critical-thinking and problem-solving skills. We have found the following to be true:

- Calculators reduce the drudgery of applying arithmetic and algebraic procedures when those procedures are not the focus of the lesson. They provide better ways to compute and manipulate symbols. For example, if the problem is to find the area of a region bounded by the graphs of two functions, then the essential challenge for the student is to understand that a definite integral is needed, determine the limits of integration, and set up the specific definite integral. Finally, the student needs to determine whether the answer obtained makes sense in the problem situation. All these tasks require serious thinking and thorough understanding. The actual computation of the integral is often best done with (or feasible only with) calculator or computer technology.
- Calculators with computer interactive geometry allow for investigations that lead to a much better understanding of geometry (Laborde 1999; Vonder Embse and Engebretsen 1996).
- Calculators help students see that mathematics has value. Students using calculators find mathematics more interesting *and* exciting. Texas Instruments first introduced a handheld calculator-based-laboratory (CBL) device in 1994 that connects to the link port of graphing calculators. This device allows students to make precise measurements of many scientific phenomena and store the measurements in their calculators for mathematical analysis. Thus, more than any other classroom innovation in the past, calculator-based laboratories have connected school mathematics to the real-world phenomena around the student. The excitement and interest in both mathematics and science generated by these real-world connections is impressive (Bruneningsen and Krawiec 1998).
- Calculators make possible a "linked multiple-representation" approach to instruction. A graphing calculator makes graphical and numerical representations practical learning strategies.

- Before calculators we studied calculus (applications of the derivative) to learn *how to obtain accurate graphs*. Today we use accurate graphs produced by a graphing calculator to help us study the concepts of calculus.

WHY THE CONTROVERSY ABOUT USING CALCULATORS IN MATHEMATICS EDUCATION?

Controversy is associated with using technology in the teaching and learning of mathematics. It is human nature not to want to change. It is comfortable to teach in the way we were taught. One of the great problems we face in mathematics education is communicating the real nature and value of mathematics. Before the publication of the *Curriculum and Evaluation Standards for School Mathematics* (National Council of Teachers of Mathematics [NCTM] 1989) and before the extensive use of calculators in mathematics classrooms, most students viewed mathematics as a bag of tricks and rules to memorize for computing or solving something. All too many students still do. Personal experience and evidence from the 1986 NAEP (Dossey et al. 1988, p. 102) strongly support this observation. Students also think of mathematics as tedious, boring work, particularly when they remember only the endless drill exercises—the "do it until it hurts" kind. We must communicate the true nature of mathematics and build a case that the appropriate use of technology will enhance the teaching and learning of mathematics. If the true nature of mathematics is understood, then the use of technology in the learning of mathematics will be seen as natural enhancements and extensions.

Paper-and-pencil arithmetic and algebraic symbol-manipulation procedures were very important in the past because they were the only procedures available for computing and solving. Today, teachers must examine on a case-by-case basis which paper-and-pencil arithmetic and algebraic manipulation procedures should still be emphasized in the curriculum. It will become clear that many techniques we teach are still emphasized in the curriculum only because they were the only methods possible in the past. We must distinguish between applying mathematics algorithms and doing real mathematics (Ralston 1999).

What Does the Research Tell Us?

We have strong evidence from careful research studies to support the use of technology in the teaching and learning of mathematics. A recent comprehensive listing and analysis of calculator research has been completed by Dunham (in press). One of the most compelling arguments for the use of calculators in mathematics teaching and learning is the meta-analysis of eighty-eight studies on the use of calculators that was conducted by Hembree and Dessart (1992). Only one of these studies reported negatively about calculator use. We cite one of the conclusions drawn from Hembree and Dessart's analysis.

> The preponderance of research evidence supports the fact that calculator use for instruction and testing enhances learning and the performance of arithmetical concepts and skills, problem solving, and attitudes of students. Further research should dwell on the best ways to implement and integrate the calculator into the mathematics curriculum. (P. 30)

The studies reviewed by Hembree and Dessart typically were conducted in classrooms in which the students were taught the traditional paper-and-pencil skills while or before they used calculators. In a longitudinal study done between 1986 and 1992 in Great Britain, the children were never taught the traditional paper-and-pencil algorithms for arithmetic, but over the years the children used calculators and successfully invented their own paper-and-pencil processes for the arithmetic operations (Shuard 1992). One of the interesting, albeit ambiguous, results of the Third International Mathematics and Science Study (TIMSS) was that internationally, on all the tests at the advanced level, students who reported using calculators in their daily coursework performed well above those who rarely or never used them (TIMSS 1998). In general, it is fair to say that the research indicates that the use of calculators does not degrade the basic skills of students. In the review conducted by Hembree and Dessart (1992), calculators did appear to have positive effects on students' problem-solving abilities and attitudes toward mathematics.

How Can We Make the Best Use of This Technology?

If we want to see both the basic skills and the problem-solving skills of our students improve in contexts that allow the regular use of calculators, then we must continue to develop methods that we might agree to call "appropriate uses," not only for calculators but for paper-and-pencil techniques as well. In this regard, we have come to the conclusion that a balanced approach of paper-and-pencil techniques and technology in the teaching and learning of mathematics is essential. We need to communicate that some traditional arithmetic and algebraic skills are still very important. Indeed, we believe they will be even more important in the future as we move to more computer-intensive learning environments. For example, without using calculators, students need to be able to multiply quickly two numbers, one of which contains a single digit or is a power of 10, and be able to divide a number by a single-digit divisor or a power of 10. Computing the products and quotients of other numbers can be left to the calculator. We want students to be able to explain why $16! = 2^{15} \cdot 3^6 \cdot 5^3 \cdot 7^2 \cdot 11 \cdot 13$, but we would not ask them to use paper and pencil to obtain the factorization.

Brolin and Björk (1992) and Wheatley and Shumway (1992) support our position. In a national project in Sweden, training in arithmetic algorithms in grades 4–6 was reduced in favor of extending the use of calculators for solving more-complicated problems (Brolin and Bjork 1992). For example, it was considered sufficient if students could divide by single digits. The results of the project showed that students who used the calculator did not lose important basic skills in algorithmic calculations. Wheatley and Shumway (1992) state that "it would be much more important that students *know when to subtract* than that they be able to use a prescribed and complex subtraction algorithm efficiently" (p. 2).

Balance means the appropriate use of paper-and-pencil *and* calculator techniques on a regular basis. Used properly, paper and pencil and calculators can complement each other. It is important to know how to estimate an answer before doing a computation using either a calculator or paper and pencil. It is important that students have enough number sense to recognize when answers are correct and that they know methods of checking answers without doing the problem over. And it is important for students to understand at least on an intuitive level why procedures work and when they are applicable. Balance does not mean that we quit teaching such skills as long division or factoring. No one should simply dictate that. However, it does mean that our objectives for mastery and understanding shift from speedy paper-and-pencil computation in division and factoring problems to making sense of the operations and their proper use.

Time must be provided in the curriculum for *appropriate* practice of these needed skills. One method teachers use to achieve a good balance is to have students routinely employ each of the following three strategies:

1. Solve problems using paper and pencil and then *support* the results using technology
2. Solve problems using technology and then *confirm* the results using paper-and-pencil techniques
3. Solve problems for which they choose whether it is most appropriate to use paper-and-pencil techniques, calculator techniques, or a combination of both

These approaches help students understand the proper use of technology.

Another approach to achieving balance is to use manipulatives and paper-and-pencil techniques during the initial concept development and use calculators in the extension and generalizing phases. There is a vast difference between achieving precision in a paper-and-pencil procedure and explaining why the paper-and-pencil procedure works. Solving systems of two linear equations in two unknowns provides a good illustration of this difference. We could expect students to solve several systems using paper-and-pencil substitution. Then using this knowledge, students could extend their understanding by solving the same systems, step-by-step, with a computer algebra system. Transferring the process to a computer algebra system requires that students understand the process. Finally, with the computer algebra system we can challenge the students to solve a general system of two linear equations in two unknowns to obtain their own version of Cramer's rule.

Whatever our methods may be for achieving the balance we are calling for, we must *not* back off on the full, regular, and integrated use of available technology, including graphing calculators with computer algebra, computer interactive geometry, and computer software microworlds, discussed below, in *all* school mathematics classes.

In order to achieve the balance we are calling for, assessment needs to be given a more prominent role in professional development activities. High school teachers being introduced to calculators with CAS often remark that if they allowed such calculators in their classrooms, they would have to change all their tests and quizzes! Professional development could address such concerns by discussing when to test with technology and when to test without it. It is all right to give some tests without technology. It is *not* all right to give all tests without technology because doing so makes technology seem unimportant and an add-on to the curriculum. The use of technology must be truly integrated into the fabric of classroom practice. Indeed, new textbooks are needed that integrate technology into the fabric of the curriculum.

Whenever students use calculator technology on a regular basis, we run the risk that some will develop misconceptions because of the limitations of the calculators or inappropriate use. For example, many instances of inappropriate calculator use stem from a lack of understanding of how a calculator draws a graph. (It samples only a *discrete* number of function values and connects the associated points.) Errors can obviously occur when a discrete device like a graphing calculator is used to model continuous functions (Demana and Waits 1988). Designers of professional development programs must understand that with every advance in calculator technology, teachers must not only be updated but also be made keenly aware of the limitations of the technology.

Teachers' fears about technology need to be understood and addressed. New CAS calculators can perform most of the traditional algebra and calculus symbolic manipulations. Unfortunately, most classroom teachers today spend the majority of their time on the very same manipulations using paper-and-pencil techniques, some of which will soon be obsolete. CAS tools do the manipulations faster and more accurately than any teacher or student. Student use of these new tools will require many changes. Curricula will change. Tests will change. Expectations will change. Teachers who are not willing to change will indeed fear technology.

At a higher level, to achieve the called-for balance, new state tests are needed that acknowledge technology. Documents like the *Principles and Standards for School Mathematics* (NCTM 2000) should be used as a catalyst for the discussion of such issues. New, generally applicable pedagogical approaches need to be developed, tested, and disseminated. For example, some Austrians have developed little-known but powerful strategies for using computer algebra systems in algebra and calculus. These strategies are known as the black-box–white-box and the scaffolding principles (Heugl, Klinger, and Lechner 1996). Consider teaching long division of polynomials. In the white-box phase no calculators would be used except perhaps to check results. Paper-and-pencil procedures would be developed that illustrate the division algorithm and why it works. Later in the year, when division is needed in a problem, students would be allowed to use a calculator for the computation (black-box phase).

RECENT ADVANCES IN CALCULATOR TECHNOLOGY

Perhaps the single most significant advance in calculator technology that has huge ramifications for the future of calculators in mathematics classrooms has been the invention of "flash ROM."

What Does Flash ROM in a Calculator Do?

Flash ROM is a new type of calculator memory first introduced by Texas Instruments in 1998. Until recently calculators had only two types of distinct memory, ROM and RAM. ROM, or read only memory, can be programmed only once and never changed. All the built-in functionality that comes with a calculator is stored in ROM. ROM is relatively inexpensive, so the amount of ROM used in calculators has increased over the years as more and more functionality has been included. If a calculator had only ROM memory, it would not be possible to enter numbers, store values into variables, or even graph a given function. For these operations, the calculator needs RAM, or random access memory, which allows new information to be stored.

RAM can be rewritten an unlimited number of times. It is used as scratch space during calculations and also as a place to store information such as equations, lists, programs, and so on. RAM has the drawback that it requires more power to operate than ROM, an important consideration for low-power, battery-operated devices like calculators. Also, RAM has the drawback of being relatively expensive. It is usually the second most expensive part of a calculator, after the display. Despite the drop in prices over the last few years for computer RAM, calculator RAM prices have not dropped as fast because calculators use a different type of RAM. To keep the price of calculators low, the amount of RAM in calculators has been restricted. Flash ROM combines the benefits of both RAM and ROM in that it is ROM but it can be rewritten like RAM, although it is currently limited to about one hundred thousand rewrites.

Flash ROM supplies much more memory in a calculator. Already, flash calculators can have six to ten times the amount of user memory found on non-flash graphing calculators. Flash ROM allows calculators to be upgraded electronically. A new version of the built-in mathematical software, or base code, can be downloaded to the calculator, replacing the previous version. Students will be able to upgrade their calculators and add the latest features without buying a new calculator. Also, calculator companies will be able to distribute maintenance upgrades that improve the underlying system without replacing the calculator itself. This feature is very important to teachers and parents for economic reasons, since it will make calculator "boxes" last longer.

Perhaps the most significant implication of flash ROM is that it enables calculator software applications, also called *flash applications,* which will allow the calculators of the future to become small computer platforms for software applications!

What Are Flash Applications?

Flash applications are software programs that run on a calculator. They can do more than user programs developed in the calculator's program editor because they are written in more-powerful software languages (C and assembly language) that tap into more of the underlying calculator system. Flash applications can also be faster than user programs for the same reason. Flash applications provide a way of adding on to the built-in functionality, or base code, with additional software that is similar in construction. Like the base code, flash applications are stored in flash ROM and remain there while running. Therefore, they do not take up valuable RAM space the way user programs do, they stay on the calculator unless they are deliberately deleted, and they can't be accidentally removed by resetting RAM or if the calculator's batteries die.

Flash applications can dramatically change the functionality of a calculator, since they are able to control what is displayed on the calculator screen down to the level of individual pixels. Flash applications are not limited to displaying the menus, home screen, tables, and graphs of a standard graphing calculator but can also display pictures, animations, icons, new types of menus, and so on. Lessons and activities that used to be delivered as worksheets or textbook exercises can be illustrated, animated, and electronically linked to the calculator's computational features.

For example, Puzzle Tanks (Sunburst Communications 1999) is a flash application for developing mathematical problem-solving skills. Puzzle Tanks animates the standard problem of obtaining a given quantity of liquid using tanks of different fixed sizes. It shows the tanks and liquid levels on the calculator display and updates them interactively as the student enters estimates (fig. 5.3). The game involves four levels of play, each level increasing with difficulty. Notice the remarkable change in the look of the traditional graphing calculator screen due to the flash application. Anyone familiar with the wide array of educational software available today can see from this example the

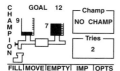 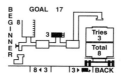 

Fig. 5.3. TI-73 screen dumps from Puzzle Tanks

educational threshold that calculators are about to cross because of flash ROM. (For other examples, see www.ti.com/calc/flash/73apps.htm#pt.)

THE MARRIAGE OF CALCULATORS AND COMPUTERS AND OTHER PREDICTIONS

No one can know with any certainty what tomorrow will bring in the area of classroom calculator technology. However, we believe our crystal ball is clear enough to make a few conjectures about the immediate horizon:

• New flash ROM calculators will hasten the current "engagement" of calculators and computers into a real marriage. Flash calculators will quickly become viewed as pocket, or handheld, computers because most computer software packages now available on desktop personal computers will be adaptable to run on tomorrow's calculators. We believe this will have the effect of tremendously accelerating the use of "computer" software by all students in schools. Just imagine the "microworlds" for teaching mathematics described by visionaries in the 1980s, and even those envisioned today, implemented on inexpensive flash calculators (see, e.g., James Kaput's SimCalc Web page, www.simcalc.umassd.edu/)! Imagine spreadsheets, three-dimensional geometry, Logo, and other powerful mathematical tools running on the calculators of tomorrow! A note of caution is in order, however. As calculators and computers assume more-prominent roles in education, integration among software applications and between software and curricula is essential for the wide-scale adoption of technology in schools (Bork 1995). Roschelle et al. (1998) point out the problem in the abstract of their paper:

> Technology-rich learning environments can accelerate and enhance core curriculum reform in science and mathematics by enabling more diverse students to learn more complex concepts with deeper understanding at a younger age. Unfortunately, today's technology research and development efforts result not in a richly integrated environment, but rather with a fragmentary collection of incompatible software application islands.

• The Internet is causing profound changes in our society and has allowed anyone to access the network using a personal computer. We believe the same changes will soon occur with networked calculators, linking them to one another, to computers, and to the Internet. This networking will have a profound effect on the classrooms of tomorrow similar to that experienced when graphing calculators brought the power of computer visualization to thousands of students who had little or no access to computers.

• We are also likely to see textbook publishers move to integrate more calculator-driven computer software into their lessons. Textbooks might even become thinner!

• The marriage of calculators and computers will allow us to resolve some of the intractable equity issues of our educational system. We predict that the inexpensive flash calculator will become the personal computer for *every* mathematics and science student and students in other disciplines, as well.

• Looking to the future, we cannot fail to note that school mathematics and science standards should be a catalyst for a discussion of the fundamental issues raised in this article. We note also that standardized tests *must* change to reflect the advances technology has made and will continue to make in the curriculum. Used unwisely, standardized tests can have a very detrimental effect on teachers' and textbook publishers' willingness to make needed changes and tackle the hard calculator issues. It will be no different in the twenty-first century unless we as a society allow our old prejudices to be put on the table for discussion.

As we conclude this article, we would be remiss not to point out a trend toward globalization that has been happening for a long time but with little fanfare. Because of the information technologies of the late twentieth century, we believe, this trend has just entered the steep part of an exponential curve of change. We all know that because of technology the world is becoming a smaller place. Innovations no longer remain dormant in isolated regions of the world. The use of graphing calculators in school mathematics is rapidly increasing worldwide. For example, they are used extensively in France, Germany, Scotland, Austria, Sweden, Denmark, Holland, Australia, Portugal, and Canada. There are now organizations like T^3 in twenty-one countries worldwide, including Japan. National tests in Sweden, Denmark, Portugal, and France *require* graphing calculators. Many countries have centralized provincial curricula and tests and require graphing calculators on these tests. In France, students are allowed to use CAS calculators on the final National Lycée exams used for university entrance. (And French high school students achieved the top score on the advanced mathematics part of the recent TIMSS study.) With the capabilities already visible on the horizon of tomorrow's calculators and with calculators' intimate connection with the other rapidly developing computer and software technologies, the impact of calculator and computer technologies on the classrooms of tomorrow will become an international issue. These classroom technologies will very much serve as a catalyst, perhaps even more than studies like TIMSS, for bringing mathematics curricula around the world closer together.

REFERENCES

Ball, Guy. "Texas Instruments Cal-Tech, World's First Pocket Electronic Calculator." 1997. www.geocities.com/SiliconValley/Park/7227/caltech.html

Bork, Alfred M. "Why Has the Computer Failed in Schools and Universities?" *Journal of Science Education and Technology* 2 (December 1995): 97–102.

Brolin, Hans, and Lars-Eric Björk. "Introducing Calculators in Swedish Schools." In *Calculators in Mathematics Education*, 1992 Yearbook of the National Council of Teachers of Mathematics, edited by James T. Fey, pp. 226–32. Reston, Va.: National Council of Teachers of Mathematics, 1992.

Bruneningsen, Chris, and Wesley Krawiec. *Exploring Physics and Mathematics with the CBL System*. Dallas, Tex.: Texas Instruments, 1998.

Demana, Franklin, and Bert K. Waits. "A Computer for *All* Students." *Mathematics Teacher* 85 (February 1992): 94–95.

———. "Graphing Calculator Intensive Calculus: A First Step in Calculus Reform for All Students." In *Proceedings of the Preparing for a New Calculus Conference*, edited by Anita Solow, pp. 96–102. Washington, D.C.: Mathematics Association of America, 1994.

———. "Pitfalls in Graphical Computation, or Why a Single Graph Isn't Enough." *College Mathematics Journal* 19 (March 1988): 177–83.

Dossey, John A., and Ina V. Mullis. "NAEP Mathematics—1990–1992: The National, Trial State, and Trend Assessments." In *Results from the Sixth Mathematics Assessment of the National Assessment of Educational Progress*, edited by Patricia Ann Kenney and Edward A. Silver, pp.17–32. Reston, Va.: National Council of Teachers of Mathematics, 1997.

Dossey, John A., Ina V. Mullis, Mary M. Lindquist, and Donald L. Chambers. *The Mathematics Report Card: Are We Measuring Up?* New York: Educational Testing Service, 1988.

Dunham, Penny. "Hand-Held Calculators in Mathematics Education: A Research Perspective." Paper presented at the NCTM Conference on Technology and Standards 2000, Arlington, Va., June 1998.

Hembree, Ray, and Donald J. Dessart. "Research on Calculators in Mathematics Education." In *Calculators in Mathematics Education*, 1992 Yearbook of the National Council of Teachers of Mathematics, edited by James T. Fey, pp. 23–32. Reston, Va.: National Council of Teachers of Mathematics, 1992.

Heugl, Helmut, Walter Klinger, and Josef Lechner. *Mathematikunterricht mit Computeralgebra-Systemen: Ein didaktisches Lehrerbuch mit Erfahrungen aus dem österreichischen DERIVE-Projekt*. Bonn, Germany: Addison-Wesley Publishing, Ltd., 1996.

Laborde, Colette. "Vers un Usage banalisé de Cabri-Géomètre avec la TI 92 en classe de seconde: Analyse des facteurs de l'intégration." In *Calculatrices symboliques et géométriques dans l'enseignement des mathématiques*, edited by Dominique Guin, pp. 79–94. Montpellier, France: Institut de Recherche sur l'Enseignement des Mathématiques, 1999.

Loucks-Horsley, Susan, Peter W. Hewson, Nancy Love, and Katherine E. Stiles. *Designing Professional Development for Teachers of Sciences and Mathematics*. Thousand Oaks, Calif.: Corwin Press, 1998.

National Council of Teachers of Mathematics. *Curriculum and Evaluation Standards for School Mathematics*. Reston, Va.: National Council of Teachers of Mathematics, 1989.

———. *Principles and Standards for School Mathematics*. Reston, Va.: National Council of Teachers of Mathematics, 2000.

Pollak, Henry O. "The Effects of Technology on the Mathematics Curriculum." In *Proceedings of the Fifth International Congress on Mathematical Education*, edited by Marjorie Canss, pp. 346–51. Boston.: Birkhauser Press, 1986.

Ralston, Anthony. "Let's Abolish Pencil-and-Paper Arithmetic." *Journal of Computers in Mathematics and Science Teaching* 18 (1999): 173–94.

Roschelle, Jeremy, Jim Kaput, Walter Stroup, and Ted M. Kahn. "Scaleable Integration of Educational Software: Exploring the Promise of Component Architectures." *Journal of Interactive Media in Education* 98 (October 1998). www.jime.open.ac.uk/98/.

Shaughnessy, Catherine A., Jennifer E. Nelson, and Norma A. Norris. *NAEP 1996 Mathematics Cross-State Data Compendium for the Grade 4 and Grade 8 Assessment*. Washington, D.C.: National Center for Education Statistics, 1998.

Shuard, Hilary. "CAN: Calculator Use in the Primary Grades in England and Wales." In *Calculators in Mathematics Education, 1992 Yearbook of the National Council of Teachers of Mathematics*, edited by James T. Fey, pp. 33–45. Reston, Va.: National Council of Teachers of Mathematics, 1992.

Sunburst Communications. "Puzzle Tanks." Pleasantville, N.Y.: Sunburst Communications, 1999. For more information, see www.sunburst.com/new_products_software.html.

Third International Mathematics and Science Study. "International Study Finds the Netherlands and Sweden Best in Mathematics and Science Literacy." Press release of the Third International Mathematics and Science Study, 24 February 1998. TIMSS.bc.edu/TIMSS1/presspop3.html.

Vonder Embse, Charles, and Arne Engebretsen. "Using Interactive Geometry Software for Right-Angle Trigonometry." *Mathematics Teacher* 89 (October 1996): 602–5.

Waits, Bert K., and Franklin Demana. "The Calculator and Computer Precalculus Project (C^2PC): What Have We Learned in Ten Years?" In *Impact of Calculators on Mathematics Instruction*, edited by George Bright, Hersholt C. Waxman, and Susan E. Williams, pp. 91–110. Lanham, Md.: University Press of America, 1994.

———. "A Computer for *All* Students—Revisited." *Mathematics Teacher* 89 (December 1996): 712–14.

———. Master Grapher (computer software). Reading, Mass.: Addison-Wesley Publishing Co., 1987.

Wheatley, Grayson H., and Richard Shumway. "The Potential for Calculators to Transform Elementary School Mathematics." In *Calculators in Mathematics Education, 1992 Yearbook of the National Council of Teachers of Mathematics*, edited by James T. Fey, pp. 1–8. Reston, Va.: National Council of Teachers of Mathematics, 1992.

6

Technology-Enriched Learning of Mathematics
Opportunities and Challenges

Frank Wattenberg

Lee L. Zia

It is easy to feel like kids in a candy store as each new year brings new and more-exciting technology for doing and learning science and mathematics—personal computers, scientific and graphing calculators, calculators and computers with computer algebra systems linked to devices for collecting scientific data, the World Wide Web, Java applets, and much more. The World Wide Web (WWW) is providing increasing access to primary resources, including expensive, site-bound, or dangerous laboratory facilities (e.g., students across the country already can use a radio telescope at the Haystack Observatory at web.haystack.mit.edu/education/education.html or experiment with liquid crystals at the ALCOM Science and Technology Center at olbers.kent.edu/alcomed/Experiment/Frames/eo_setup.html); to massive and real-time data sets; and to museum-quality collections. Other advances in technology open up more possibilities—for example, consumer-level Postscript printers can produce finely detailed slides that can be used with inexpensive (under $20) laser pointers to experiment with diffraction and to understand the mathematics behind the discovery of the structure of DNA (curriculum materials and details are available from the Institute for Chemical Education (ICE) at jchemed.chem.wisc.edu/ice/reference/Optical_Transform.htm or

---

An expanded and interactive version of this paper is available on the World Wide Web at www.math.montana.edu/~frankw/ccp/talks/nctm.htm. Ideally, it should be used with a Postscript printer, a Texas Instruments graphing calculator and CBL, and a laser pointer at hand.

jchemed.chem.wisc.edu/ice/). Mathematics with its tools for analysis and visualization is at the very center of the best technology-enriched, inquiry-driven learning, and mathematical modeling is especially important to enable our students to understand and build the sophisticated and compelling simulations that are becoming so common and important.

This plethora of possibilities raises questions: "What tools have been shown to improve our students' learning?" and "When and how do we use them?" This article offers an extended example of the use of some of these tools and examines some of the issues and trade-offs that mathematics teachers must consider. Finally, we speculate briefly on a digital library to support the distributed development, discovery, use, assessment, and validation of tools that will help teachers and students as they cross the traditional temporal and spatial borders of the classroom in their pursuit of science and mathematics learning.

THE MATHEMATICS OF WAVES: A THEME

Waves are all around us—visible waves like the waves in an ocean or lake and invisible waves like sound waves and light waves. Waves are a great source of delight and surprise for students learning mathematics. In this section we show how waves can be used as a theme for unifying the study of science and mathematics.

Shadows (Light) and Waves

It is easy to do some simple experiments exploring light and shadows and at the same time exercise a bit of geometry using a flashlight and some patterns printed on paper. Make a slide by drawing a picture on transparency film. Using a small light source, project your slide onto a wall as shown in figure 6.1. Do the following experiments:

- Keeping the position of the slide fixed, move the bulb toward and then away from the slide. What happens to the shadow on the wall?
- Measure the distance between two points on the slide. Measure the distance between the corresponding points in the shadow on the wall. Repeat this for several different pairs of points. Do you notice any pattern?
- Measure the distance from the bulb to the slide and from the bulb to the wall.

These experiments lead to some general observations. When the

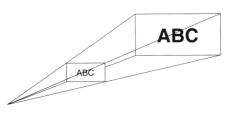

Fig. 6.1. Projecting a slide

bulb is moved toward the slide, the shadow gets bigger; when it is moved away from the slide, the shadow gets smaller. Also, the shadows of points that are farther apart on the slide are farther apart on the wall. This is a good setting in which students can see the power of geometry and ratios.

The next set of experiments, based, in part, on a talk given by Arthur Ellis, of the University of Wisconsin—Madison, look at the "shadows" produced by a laser pointer and a slide with finely ruled lines. You can obtain the necessary slides electronically from a library of graph paper and other patterns (visit www.math.montana.edu/~frankw/ccp/GraphPaper/diffraction/index.htm and click on the link for "Twelve copies of a three-way grid"; an even better slide can be obtained from the Institute for Chemical Education [ICE] at jchemed.chem.wisc.edu/ice/reference/Optical_Transform.html or jchemed.chem.wisc.edu/ice/) at the Connected Curriculum Project, located at www.math.montana.edu/~frankw/ccp/home.htm. Print the file you obtain on transparency film using a laser printer. Each sheet will have twelve copies of a pattern like the one shown in figure 6.2. Cut the copies apart, giving one to each student. The slides can be used with a laser pointer in two ways:

- Shine the laser pointer on a lightly colored wall. Have the students look through the slides at the dot on the wall made by the laser pointer. Caution them never to look directly into a laser pointer. They will see different patterns depending on which part of the slide they look through.
- Shine the laser pointer at a lightly colored wall through different parts of the slide and look at the "shadow" produced on the wall.

Next, do experiments with the laser pointer and three-way slides similar to the ones done earlier with an ordinary bulb and hand-drawn slide. You and your students may be surprised at the results:

- Moving the laser pointer closer to or farther away from the slide does not magnify or reduce the "shadow" on the wall.
- Lines that are *closer together* on the three-way slide produce "shadows" that are *farther apart* on the wall.

Geometry and ratios do a good job of describing the shadows produced by an ordinary lightbulb and most ordinary slides, but the patterns produced by a laser pointer and a slide with finely spaced lines are very different. We can

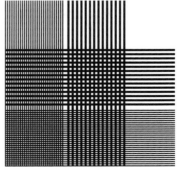

Fig. 6.2. Three-way grid

begin to explain our observations in this setting using a mathematical analogy to waves.

Ideally, one would experiment with real waves using a ripple tank. But if this is not possible, you might try a computer simulation of a ripple tank. Figure 6.3 shows a snapshot of a Java-based simulation that is available on the Web. (This "physlet" is part of the WebPhysics project at webphysics.davidson.edu/welcome.htm. The URL for the physlet collection is webphysics.davidson.edu/Applets/Applets.html.) With this "live" physlet you and your students can investigate the interference patterns produced by two sources as the distance between them is varied: when the sources are closer together, the interference pattern spreads out. This is a clue that the phenomena we observed with the laser pointer may be analogous to phenomena involving water waves. Indeed, students can revisit the idea of wave interference when they study functions like $f(x) = \sin(\omega x) + \sin(\omega x + \delta)$ in trigonometry class and even beyond in calculus, differential equations, and partial differential equations classes.

Fig. 6.3. Ripple tank physlet

You don't even need a real or computer-based ripple tank to study interference. The same library that contains the three-way grid slides used above also contains a file that will print concentric circles like those in figure 6.4, only finer. (Visit www.math.montana.edu/~frankw/ccp/GraphPaper/index.htm and click on the link "Patterns that can be printed on transparency film and used for ripple tank experiments." Another possibility is to photocopy fig. 6.4 directly from this paper onto a sheet of transparency film.) Make copies on plain paper and transparency film. Then lay the transparency film pattern on top of the plain paper pattern to produce interference patterns. Varying the distance between the two centers (sources) leads to results similar to those from a real or virtual ripple tank. This low-tech apparatus introduces a tactile element and gives students something they can carry home.

Fig. 6.4. A low-budget ripple tank—waves emanating from a point source

Sounds and Waves

We can "look" at sound waves using a soda bottle and the Calculator-Based Laboratory (CBL) produced by Texas Instruments (TI). (The TI-CBL can be used with the TI-83, 85, 86, 89, 92, or the newer "plus" versions. In this paper we use it with the TI-92, but all the experiments we describe can be done equally well with the other TI graphing calculators.) Figure 6.5 shows a TI-CBL connected to a TI-92 and a "microphone," which really measures the variations in air pressure that we call "sound." The screen snapshots in figure 6.6 show the results of two experiments using this apparatus to record the sounds (239 air pressure readings spaced 0.0001 seconds apart) produced by

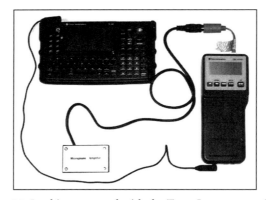

Fig. 6.5. Looking at sound with the Texas Instruments CBL

blowing across the mouth of a soda bottle. In one figure the soda bottle was empty, and in the other it had some water in the bottom. Which is which? (Hint: On the right in fig. 6.6 we see five full periods. Since the sound was recorded for 0.0239 seconds, each period is roughly 0.0048 seconds, so the frequency is roughly 208 Hz.) Students may perform experiments like these using bottles of different sizes and shapes with different amounts of water in

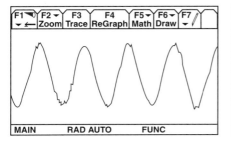 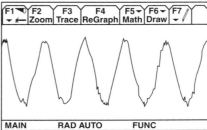

Fig. 6.6. The sounds of soda bottles

them to investigate how the frequency is related to different characteristics of the bottles. This illustrates the kind of open-ended experimentation made possible by today's low-cost and flexible scientific equipment.

For our next experiment a constant tone of 440 Hz was played through a single speaker. The graphs shown in figure 6.7 were made using two microphones attached to the same CBL, with one microphone slightly farther away from the sound source than the other one. There are two notable differences between the two graphs:

- The amplitude of the signal recorded by the microphone that is farther away from the source is lower than the amplitude of the signal recorded by the microphone that is closer to the source.
- The signal recorded by the microphone that is farther away from the source lags behind (that is, its graph is shifted to the right) the signal recorded by the microphone that is closer to the source.

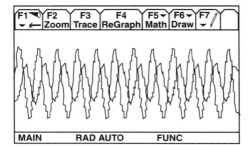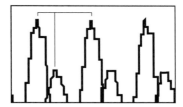

Fig. 6.7. Recordings make by two microphones, with a magnified view on the right

Several characteristics of sound can be discovered from this figure and similar ones made by varying the placement of the two microphones. For example:

- Students can determine the frequency of the signal. There are about 11.6 cycles shown in figure 6.7. Since this graph corresponds to a recording time of 0.0239 second, the duration of each cycle is roughly 0.0021 second. In other words, the period is roughly 0.0021 second and the frequency is about 1/.0021, or 476, cycles per second. This is a rough estimate (recall the tone played was 440 Hz).
- The right side of figure 6.7 shows a magnified portion of the left side. Looking at this figure, we see that the signal arriving at the microphone that is farther away from the source lags about one-third of a period, or 0.0007 second, behind the signal arriving at the closer microphone. The two microphones were 14 inches apart. Thus the speed of sound is approximately 20 000 inches, or 1666 feet, per second. This is a very rough estimate (the actual value is closer to 1200 feet per second).

In situations like this, discussions of error estimates and nuance in measurement open up rich and interesting mathematical investigations for students.

Finally, we can duplicate the ripple tank experiment with sound waves instead of water waves. This experiment should be done outside in an open area to avoid sound reflections. Place two speakers on the ground two or three feet apart and play a cassette recording of a 440-Hz tone. If you walk parallel to the line through the two speakers, perhaps fifteen feet away, you will notice there are places at which the sound "fades out." You are hearing the same interference phenomena you saw before. Such an experiment provides a visceral knowledge that complements and reinforces students' mathematical knowledge.

TOOLS: CURRENT AND FUTURE

The extended example above illustrates the variety of tools already available to the classroom mathematician. We interpret the term *tools* broadly to include both the underlying physical or virtual devices and the lessons or explorations that use them. Within this changing tool space, we shall next describe several representative types of tools and speculate on the direction in which they might evolve.

SIMPLE PHYSICAL INSTRUMENTS OR "MACHINES" COUPLED WITH SENSING DEVICES

Low-cost probes or sensors together with data-acquisition devices such as calculator- and microcomputer-based laboratories have already had a profound though perhaps not widespread effect on classroom practice, particularly in secondary school settings. The sound-wave experiments above provide concrete examples. Such devices have changed the "rule of three" to the "rule of four," adding *tactile* investigations to numerical, graphical, and symbolic investigations. In the future these hybrid analog-digital devices will be smaller with more functionality and analytical power. These developments will permit devices to share sensory ability, data-gathering functions, and data dissemination and enable the assembly of more sophisticated devices from modular building blocks.

Worldware Products

Many of these types of tools exist, mostly commercial off-the-shelf products: spreadsheets; several computer algebra systems (CAS), each with its fervent champions; and more-specialized products, such as Geometer's Sketchpad or scientific and statistical computing packages like Matlab or S-Plus. As these products gain use at lower and lower grades, vendors will create modularized "scaled down" versions designed specifically for education, possibly keyed to developmental levels.

Network-delivered applications based on Java or other languages independent of platform and operating systems will also have a profound impact—for example, Java-based spreadsheets (see www.tidestone.com or www.vni.com) are already available. The physical instruments and sensing devices described above will interoperate with these network-based tools. Virtual reality simulation and animation packages will be standard fare based on the developing Virtual Reality Modeling Language (VRML) and its successors. Haptic or force-feedback devices (e.g., joysticks that push back on the user) are being used now by game players and have great potential for science and mathematics learning. We may even see affordable holographic displays. Imagine the powerful effect on student learning of a kinesthetic interaction with the graph of a surface.

Programs, Scripts, or Macros Based on Existing Worldware Products

Current examples include, among others, spreadsheet macros, Matlab scripts, and VRML worlds. In the vast majority of instances these tools have been created in relative isolation—a situation that has led to a proliferation of many functionally similar programs, overlapping in their core computational or graphical display objectives but exhibiting variations in operational control, output representation, and choice of notation. Ironically, continual vendor upgrades of the underlying worldware products have themselves contributed to this state of affairs, since it is not uncommon for these upgrades to cause incompatibilities among versions of the same macro. Thus, despite a common core of learning objectives, many of these special-purpose programs cannot be shared beyond the local laboratory or classroom. Overcoming this inefficiency in the development and sharing of resources will be an important activity for the community of classroom mathematicians.

Structured Lessons Based on Worldware Products

The current incarnation of these lessons is an electronic notebook that offers the learner a structured exploration of a mathematical concept or possibly even of an entire topic. For example, large amounts of material for learning calculus have been developed along these lines. Depending on the capabilities of the underlying worldware product, the notebook may feature (in addition to plain text) graphics, animations, and even audio, all of which can be controlled through input forms like slider bars, buttons, and check boxes. These features permit varying degrees of interactivity for the learner and control of initial configurations through customizable or changeable parameter values. With the explosive growth of the World Wide Web, we will see a dramatic increase in the availability of these features in hypermedia formats that interact with a worldware computational or display engine as a back-end. Interoperability with handheld physical devices will also be realized, along with advances in input-output mechanisms supporting voice control and other steering capabilities.

Homegrown Software Utilities

In the early days of computing, only individuals with particular programming expertise could develop software tools for learning. Today we have stunning symbolic and numeric computational power, highly sophisticated graphical capabilities, and state-of-the-art communication features on our desktops and in our palms at affordable prices. As design, writing, and production tools mature and as human-computer interfaces improve, the pool of potential developers (teachers *and* students) will widen. Through the network, learning tools and environments will be developed, shared, and assessed collectively, increasing their effectiveness and usage.

GUIDEPOSTS FOR CHOICES

With respect to the technology-rich networked learning tools of the twenty-first century, choice is the dominant issue facing the practicing mathematics teacher. Three principal factors will influence the choices to be made: *discovery, assessment,* and *reliability.* Discovery precedes the opportunity to choose; assessment establishes a basis for choice; and finally, stability and reliability influence the decision to choose again.

Discovery

In early 1999 a request submitted to one of the standard WWW search engines for the keyword *polynomial* returned a mere 39 014 035 hits. A slightly more refined search for the key phrase *quadratic formula* returned 2 184 459 URLs, the term *Euclidean geometry* produced 853 449 hits, and *centroid* produced 7 707 possible links. The "top" (highest rated) link in this last list had the following descriptor:

> "What is a Centroid? A centroid can be thought of as a simple inverted index mechanism...to provide hints as to the location of data in a large, loosely coupled distributed database. ..." 82% Date: 27 Jul 1998, Size 7.9K weeble.lut.ac.uk/Reports/Centroid-use-in-ROADS/node2.html

Although the Web has enabled many more resources to be "published" (*posted* is more accurate), the discovery of usable resources remains difficult.

Anecdotal evidence suggests this issue is central to why teachers have still not wholeheartedly embraced the Web as a resource. *There is a pressing need for reliable tools for discovery.* This is an active research area among computer and information scientists. (See the Digital Library Initiatives sponsored by NSF, NASA, DARPA, NLM, NEH, the FBI, and others at dli.grainger.uiuc.edu/national.htm.)

Future developments will include more sophisticated refinable searches (in use now at a basic level), natural language searching, and nontextual searching over sounds and images (including mathematical formulas). Robust, accurate, and above all useful search engines will rely on new collection, cataloging, and indexing capabilities. Crucial to these tasks will be the use of metadata, for which the user community will need to help develop an appropriate descriptive vocabulary. The digital equivalent of ISBNs will enable content to be associated with user reports and other reviews so that search mechanisms can return reviews with content and even prioritize hits on the basis of reviews. Search engines will also exploit user profiles ("I speak Spanish" or "I know calculus") to find appropriate resources.

Assessment

To form a basis for their choice of learning tools, practicing teachers should have foremost in their minds an understanding of what learning goals the tool is designed to promote, and they will need evidence of the tool's effectiveness in achieving those goals. Beyond addressing student learning, the tools that thrive in the future will demonstrate their potential usability by exhibiting several additional and essential characteristics:

- *Authenticity*—Does the tool do what it is advertised to do? Is there a measure of the quality of the data that might accompany the tool?
- *Interactivity*—Does the tool permit the learner to change parameters, to enter one's own equations or otherwise define a scenario, to select among a variety of displays—symbolic, numerical, graphical, textual, even auditory?
- *Interoperability*—Does the tool share data and data analysis results conveniently?
- *Installability*—Is the tool easy to download or otherwise install? Does it have a minimal start-up time, or at least a high ratio of perceived value to start-up time?
- *Customizability*—Can the tool be easily modified to suit the needs of a broad range of learners or the demands of a broad range of learning environments?
- *Reusability and extensibility*—Can the tool be combined with other tools to create new materials? Is it reusable in other contexts, particularly multidisciplinary ones?

Reliability and Stability

As tools and learning resources grow increasingly reliant on complex hardware, operating systems, software, and delivery by the network, virtual shelf life takes on increasing importance. (Readers of this volume will survive a broken link to a URL cited here, but they will be much less satisfied if "Access is denied" or "File is not found" appears when their students attempt to link to a Java applet.) The tools that thrive in the future will demonstrate (advertise?) reliability in several additional value-added ways:

- Through their documentation not only of the basic "how to" of a tool but, more important, of implementation strategies and suggestions for extensions and other applications, plus links to other data sources
- Through their degree of platform independence, and if platform dependent, then through the provision at the very least of specifications that ensure cross-platform capability
- Through their provision of user support
- Through the quantity and quality of their "living" annotation—that is, their usage histories and other forms of review—supplied by the networked user community

TRADE-OFFS

Two particular trade-offs face the classroom mathematics teacher in the technology-enriched, network-enriched learning environment of the twenty-first century: (1) *computer-based simulated experimentation versus real, hands-on experimentation;* and (2) *the use of dedicated tools such as Java applets versus general-purpose tools such as computer algebra systems.*

Computer-Based Simulation versus Hands-On Experimentation

Experimentation in virtual labs is immensely attractive, but this must be balanced by experience with real labs and an understanding of the mathematical modeling that underlies virtual labs. Simple handheld devices such as the CBL apparatus and network-delivered materials such as the concentric circle interference pattern generator (see the discussion about fig. 6.4) should be used alongside the virtual ripple tank. Mathematics teachers have a particular responsibility and opportunity in this regard. As simulations become more common, students must have the mathematical background and skills needed to understand existing simulations and construct new ones. Above all, they must have a healthy understanding of the differences between (*a*) simulations and controlled experiments and (*b*) observations of phenomena in their natural setting.

Dedicated Tools versus General-Purpose Tools

Although the ripple tank physlet (see fig. 6.3) is both striking and easy to use, there is a danger that students may view it as a black box, appreciating the pictures without understanding their meaning. Moreover, its design is limited—in particular, it does not display the decrease in amplitude of the waves as they move away from their source. This is especially relevant if physical experiments with sound are done where the wavelength has the same order of magnitude as the distances involved.

In contrast, figure 6.8 shows a single frame of a compelling animation produced by a few lines of Mathematica code. (It could also have been produced with any CAS or with most graphics packages.) Not only is the "three dimensional" output better than the ripple tank physlet, the extra student time required to produce such pictures is repaid by improved general graphics skills, by increased understanding, and by the flexibility to pursue one's own inquiries outside the built-in options of the tool.

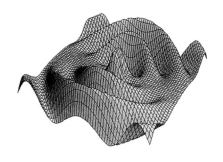

Fig. 6.8. One frame from a Mathematica ripple tank simulation

Special-purpose or dedicated software is attractive for several reasons: start-up time is relatively short, and graphical interfaces illustrate specific concepts. It is at least equally important, however, for students to develop the skills needed to use general-purpose software effectively. This necessity furnishes new opportunities for teachers to reengineer learning opportunities and their assessment of learning. For example, the analysis of a student-designed Web-based module or some other form of an electronic notebook could yield insight into the student's understanding. Many teachers have already been struck by this phenomenon when they have added open-ended writing assignments to multiple-choice and other short-answer exercises.

A Digital Library for Science and Mathematics Education

How can we harness the vast potential of technology to enable our students to learn more—and more effectively? In its short lifetime, the Web has fostered a networked community of learners (teachers and students alike), with enormous potential to influence the development, identification, assessment, and distribution of high-quality tools and other resources to improve students' learning of science and mathematics. But in its relatively youthful state the Web also exhibits serious shortcomings. It is difficult to discover high-quality and appropriate resources; resources lack reliability and stability; interoperability is more promise than reality; and progress is often hampered by a lack of reusability. A national digital library for science and mathematics education might address these shortcomings and help us realize the true potential of all this technology by providing—

- collections of high-quality, interoperable, reusable, reliable, and stable resources for grades K–16 science and mathematics learning;
- sophisticated portals or gateways to resources in its distributed collections;
- capabilities for intelligent discovery through effective indexing, abstracting, metadata tagging, and linking that support the robust search and retrieval of material;
- forums for the submission, review, and recognition of newly developed science and mathematics learning tools for use in grades K–16 and for ongoing commentary regarding their use.

This concept has already been developed in part by a series of meetings and projects supported by the National Science Foundation. Links to this work can be found at www.dli2.nsf.gov/addendum.html along with more information about NSF support for related work. (See the report of the Science, Mathematics, Engineering, and Technology Education Library Workshop at www.dlib.org/smete/public/report.html in particular.)

Conclusion and a Look Ahead

In this paper we have considered a number of different types of technology-based mathematics learning tools and speculated on the directions in which they might evolve. Although we are confident that some of these speculations will come to pass, perhaps the only sure prediction is that the number and variety of tools and the range of their applicability will increase at a rate far greater than any individual teacher will be able to follow. The collective power of a networked user community holds great promise for managing this impending overflow of data in a way that will produce real, usable information to guide the choices that teachers will make.

What additional challenges lie ahead? From a programmatic viewpoint, preservice teacher preparation and in-service teacher development programs must change with advances in technology. Teachers must be able to use the unforeseen technologies of the future. Similarly, implications will arise for certification and licensure. The form that current learning environments take is already being challenged (e.g., by distance learning), and future technological innovation will furnish even more options. Socioeconomic perspectives are also important, since issues of access and equity must not be ignored. Legal and economic restrictions on easy access to, reproduction of, and use of Web content will have a profound impact on whether real progress can be made toward eliminating duplication of effort and whether the promise of multidisciplinary perspectives and collaboration can be realized.

Finally, from a technological standpoint, dramatic increases in bandwidth and transmission speed will enable vast improvements in visualizations and animations. For example, we envision physical "walk throughs" of surfaces or trajectories. Similarly, interactive "steering" through large data sets may be possible. Expanded communication capabilities will also enable the creation of sites for sharing and exchanging that are simultaneously real for some participants, virtual for others, and possibly asynchronous for still others. New human-computer interface designs, including haptic devices and other input-output devices, will permit radically different modes of interaction with mathematical objects and mathematical models.

The twenty-first century will be nothing if not exciting. Hop on board!

ADDITIONAL RESOURCES

Web Resources

An enhanced online version of this paper can be found at www.math.montana.edu/~frankw/ccp/talks/nctm.htm.

The **Haystack Observatory Undergraduate Educational Initiative** provides remote access to a radio telescope at web.haystack.mit.edu/education/education.html.

The **ALCOM Education Project** of the **ALCOM Science and Technology Center** provides remote access to experimental equipment for liquid crystals at olbers.kent.edu/alcomed/dhtml1.html.

Material by **Arthur Ellis** in the Chemistry Department of the University of Wisconsin—Madison and others on diffraction and other subjects may be found through the **Institute for Chemical Education** (ICE) at jchemed.chem.wisc.edu/ice/.

The **WebPhysics** project, which was started in spring 1995 by **Wolfgang Christian** (Davidson College) and **Gregor Novak** (IUPUI), is located at webphysics.davidson.edu/ and is a wonderful source of applets in physics.

The idea of a National Digital Library for Science Education has been developed in a number of reports supported by the National Science Foundation, including the following:

- Report of the Science, Mathematics, Engineering, and Technology Education Library Workshop held 21–23 July 1998 can be found at www.dlib.org/smete/public/report.html.

- Serving the Needs of Pre-College Science and Mathematics Education: Impact of a Digital National Library on Teacher Education and Practice:

Proceedings from a National Research Council Workshop held September 1998 is at www.nap.edu/catalog/9584.html.
- Developing a Digital National Library for Undergraduate Science, Mathematics, Engineering, and Technology Education: A Report of a National Research Council Workshop, 7–8 August 1997 is at books.nap.edu/books/0309059771/html/R1.html.
- Report from Digital Libraries and Education Working Meeting, held at the National Science Foundation on 4–6 January 1999, is found at www.dli2.nsf.gov/dljanmtg.pdf.
- Additional information can be found through www.dli2.nsf.gov/addendum.html.
- Information on the Digital Libraries Initiatives can be found at dli.grainger.uiuc.edu/national.htm.
- The NSF workshop on Information Technology: Its Impact on Undergraduate Education in Science, Mathematics, Engineering, and Technology (NSF 98–82) is at www.nsf.gov/cgi-bin/getpub?nsf9882.
- A National Digital Library for Science, Mathematics, Engineering, and Technology Education by Frank Wattenberg, D-Lib magazine, October 1998 www.dlib.org/dlib/october98/wattenberg/10wattenberg.html

Print Resources

Brand, Stewart. *The Media Lab: Inventing the Future at MIT*. New York: Penguin Books, 1988.

Committee on Issues in Transborder Flow of Scientific Data, National Academy Press. *Bits of Power: Issues in Global Access to Scientific Data*. Washington, D.C.: National Academy Press, 1997.

Fortenberry, Norman L., and Frank Wattenberg. "Planning for a Digital Library for Science, Mathematics, Engineering, and Technology Education." In *Computers and Advanced Technology in Education*, edited by M. H. Hamza, pp. 3–7. Calgary, Alta.: ACTA Press, 1999.

Lesk, Michael. *Practical Digital Libraries: Books, Bytes, and Bucks*. San Diego, Calif.: Academic Press/Morgan Kaufmann, 1997.

Negroponte, Nicholas. *Being Digital*. New York: Vintage Books, 1995.

NII 2000 Steering Committee, National Research Council, National Academy Press. *The Unpredictable Certainty: Information Infrastructure through 2000*. Washington, D.C.: National Academy Press, 1996.

———. *The Unpredictable Certainty: White Papers*. Washington, D.C.: National Academy Press, 1998.

Standage, Tom. *The Victorian Internet*. New York: Walker & Co., 1998.

7

Using Extranets in Fostering International Communities of Mathematical Inquiry

Lyn D. English

Donald H. Cudmore

T<small>HE</small> problem shown in figure 7.1 is one of many that have been created by grade 8 and 9 students in an international mathematics program we have been implementing since 1997. Schools from a number of nations, including the United Kingdom, Canada, Taiwan, and Australia, have been participating. Our primary aim has been to establish cross-cultural communities of mathematical inquiry where students participate in shared data-handling investigations that have a major problem-posing component. Such communities enable all students to be actively involved in building meaningful mathematics.

Hilda's Bedtime Problem

One fine morning, Hilda woke up after a long night's sleep (going to bed at 11:00 pm) and began her daily routine. First to look in the mirror. 'AHHHHHH-HHHH' she screammed [sic], as the bags under her eyes horrified her, 'Am I going to bed too early, too late?' she questioned, 'I must find out.' Answer the following questions to help solve Hilda's problem:

> *1.)What is the overall average, range, and mode bedtime (weeknight) hours for Canadian and Australian students?*
>
> *2.)What time should Hilda go to bed? Which is more acurate [sic] - average or mode ?*

(created by Jessica, grade 9)

Fig. 7.1

The focal point for our activities and interactions is an "extranet" on the World Wide Web. An extranet has two principal features, namely, the ability to (1) display Web pages, which may be password protected, to authorized individuals anywhere in the world, and (2) automatically accept, process, and distribute data using Web-based forms. The words of Pfaffenberger (1998, p. ix), a business writer, are worth citing here:

> Discounting the usual hype that occurs when the "next big thing" in computing comes along, Extranets may represent the most significant development in enterprise computing since the desktop PC.

Extranets can be very complex and commercial, such as an Internet banking site, or very simple, such as a Web page that contains an electronic guest book for visitors to sign. At one time, a substantial amount of computing knowledge was required to create and manage extranets (Pfaffenberger 1998; Bernard 1998). However, at this writing, we have seen the emergence of a new breed of World Wide Web sites that enable nonprogrammers to readily build and host their own extranets cost-free. Schoollife.net (www.schoollife.net), for example, provides an extranet-builder—which it refers to as an online "community publishing system"—that is targeted directly at schools, educators, and students.

As educators, we are particularly interested in the use of interactive Web-based extranets to promote the exchange of information both within and among classrooms around the globe. Software to support such interaction is often described as "groupware," especially if it allows some measure of privacy for those who are exchanging the information (Pfaffenberger 1998).

For mathematics education, extranets can open up new learning opportunities for students and teachers alike. For example, they can provide an environment that fosters the sharing of mathematical questions, data, ideas, and student-generated problems among students locally, nationally, and internationally. Students' mathematical experiences are no longer confined to their classroom; students can take their mathematical understandings into unexplored territories (e.g., in responding to a novel mathematical investigation by a student group in another country). Likewise, an extranet enables teachers to engage readily in cross-cultural curriculum discussions, to develop and refine collaborative projects, to share feedback on their students' progress on these projects, and to give one another encouragement and support in implementing new ideas. Indeed, we consider extranet-based communities of inquiry to be one of the many exciting and powerful means by which our students and teachers will engage in mathematics in the twenty-first century.

The challenge, of course, is getting these communities started and encouraging their growth. In meeting this challenge, we have been establishing learning environments that try to maximize the educational features of an extranet.

COMMUNITIES OF MATHEMATICAL INQUIRY

The idea that teaching and learning are inquiry-based activities is not new. The National Council of Teachers of Mathematics (NCTM) has been emphasizing this point for many years (NCTM 1989) and continues to highlight it in its recent *Principles and Standards for School Mathematics* (2000). Of importance is the call for a restructure of the curriculum to focus on the engagement and exploration of problem situations where students' thinking is valued (Romberg and Collins in press). This approach has been implemented in several instructional programs designed to establish communities of learners (e.g., the program Fostering Communities of Learners, developed over the last decade by Ann Brown and her colleagues [Brown and Campione 1996]).

Central to our learning community is the mathematical inquiry process, where participants (students, teachers, and university educators) are motivated to explore mathematical situations that are intriguing, problematic, challenging, and inviting. Such situations include, among others, data handling and statistical problem posing, which we address here, and model-eliciting activities in which students produce a model that describes certain relationships, patterns, and operations inherent in a real-life situation (Lesh et al. in press). As well as developing important mathematical concepts, model-eliciting activities require students to externalize their thinking and reasoning by describing, explaining, constructing, modifying, and refining their understandings and viewpoints. Computer-supported environments, especially networked environments, furnish unique opportunities for students to display and develop these processes. For example, in one activity that we have implemented, students researched and analyzed reports on the costs of living today and in bygone times. The students then developed models that presented a case on whether or not twelve-year-olds have a better standard of living today. In sharing their models with their international peers, the students gained firsthand knowledge of current consumer costs in other nations. Such exchanges enriched both students' mathematical understandings (e.g., applying proportional ideas to new situations) and their appreciation of life in other cultures.

The nature of the communication among participants is a principal factor in building communities of inquiry, especially when these communities span several continents. In our own work, we try to extend students beyond basic discourse to include a focus on a range of thinking and reasoning processes that might assist them in constructing, analyzing, and conveying mathematical ideas. Table 7.1 presents some examples of the processes we address. Not only is communication of this extended kind one of the main goals of curriculum reform (Silver 1996), but it also is becoming increasingly important as students become more and more immersed in new technologies.

TABLE 7.1
Reasoning Processes for a Community of Mathematical Inquiry

- Analyzing concepts and identifying relationships
- Drawing distinctions and connections among ideas
- Reasoning by analogy
- Reasoning deductively and inductively
- Thinking in diverse ways, including critically, creatively, and flexibly
- Exploring alternatives and different possibilities
- Taking all relevant considerations into account
- Formulating and applying criteria
- Constructing explanations and reasons, and distinguishing between effective and ineffective examples
- Posing and critically evaluating mathematical problems and questions
- Understanding and evaluating arguments
- Anticipating, predicting, and exploring consequences
- Constructing inferences
- Generating and testing hypotheses
- Acknowledging and respecting different perspectives and viewpoints
- Developing a commitment to the processes of inquiry and their improvement, including one's own thinking and reasoning processes

To illustrate how we try to nurture these communities, we reproduce here a discussion that took place with some ninth-grade students early in one of our data-handling programs. The students' and teacher's comments refer to a newspaper article titled "First-Time Parents Nurse a Boy Bias." The article opened with the statement, "More first-time parents dream of having a son than a daughter." The reported study involved twenty couples aged between twenty-five and forty-three. It was found that eight mothers and eleven fathers preferred a boy, whereas only five mothers and two fathers preferred a girl. The remaining parents indicated no preference.

Teacher: Now that we've read the article, what do you think about the opening line? Is it an appropriate one? What's your opinion, Jodie?

Jodie: Yeah, I think it's OK because the evidence, you know, pretty much proved that.

Mackenzie: What do you mean by the evidence, like the evidence in the article?

Jodie: Yeah, just in this article. Yeah.

Anita: Yes, I think it is appropriate to say that, because that's sort of what the whole article's about.

Teacher: Does anyone have other views on this?

Catherine: Well, I don't think their statement is appropriate because it's only—they've only interviewed twenty couples, and they haven't actually said they've only interviewed twenty. It's not very accurate.

Teacher: Well, what would you do if you wanted to make it more accurate?

Catherine: I'd say, "Out of the twenty couples we interviewed." I wouldn't say, "More first-time parents dream of having a son than a daughter," because only twenty couples were interviewed, and they should have said, "More first-time parents *out of the twenty couples we interviewed.*"

Teacher: If you were to do some research to find out whether there's a bias in what first-time parents prefer, how would you go about it?

Catherine: Well, um, I'd probably interview more than twenty.

Teacher: So, based on this research they've reported, would you accept their conclusions?

Mackenzie: No, not on that research, because they haven't interviewed enough people.

Teacher: OK, so we have to interview more people, but what else should we consider ... like, what sort of first-time parents might have been interviewed?

Catherine: I think just your average person—twenty-five-to-forty-three-year-olds. There're no really young ones. And no old ones.

Teacher: And how does this affect things?

Anita: Well, when you're younger, I suppose, we don't even know if they want a child if they're sixteen.

Catherine: They should have interviewed more, and you know, gone to hospitals, and ... also, interviewed people who can't have children and who adopt them. They should have interviewed bits and pieces of everybody—people who aren't considered, you know, average—sort of different beliefs.

Teacher: What sort of different beliefs?

Catherine: Different religions, different tastes and things.

It is difficult to foster these communities without an atmosphere of trust and respect for participants and their different ideas. That is, participants should have the confidence and trust in their community membership to express their thoughts and feelings about mathematics and about issues that arise from their investigations. They should feel free to ask questions that matter to them and know that others will listen to their questions and respect their viewpoints. These aspects are essential to a productive international community.

We now consider one way of establishing an international community by an extranet—a community that reflects the spirit of mathematical inquiry. We first highlight some practical issues that warrant consideration in using extranet technology.

USING AN EXTRANET

The first issue in using an extranet to foster local, national, and international communities pertains to joining or establishing a suitable Web site. We will not address the numerous extranet sites that can be joined through the Internet or the Web sites that offer models for establishing Internet collaborations. Rather, we identify a few important questions that need to be considered in setting up an extranet link with other classrooms:

- What types of information will your site display? Who will display them, and to whom?
- What types of information will your site process, and how will they be processed?
- How will security be handled, that is, which parts of your site will be public domain and which will be semiprivate (accessed through a password)?
- How will appropriate conduct in publishing material be ensured?

How we addressed these questions in our own work is described in the next section.

The second issue to consider is how an extranet can support a mathematical community beyond the classroom. Such a community can comprise interactions among students, among teachers, among university educators, and among teachers and university educators. We have structured our extranet site to enable all these interactions to take place through specifically allocated zones that we refer to as "forums." For example, Discussion Forums enable each student in each class to share and build on the mathematical ideas of his or her peers, whether they be local, across the country, or around the world. Likewise, through our own Discussion Forums, teachers and university researchers share classroom observations, plan collaborative learning activities, discuss and monitor students' progress, and assess and refine the overall program. Our public-domain home page acts as a gateway to these forums and to other zones on the site (a more detailed discussion on the features of our extranet is given in Cudmore and English [1998]).

Mathematical Inquiry through an Extranet: Data Exploration and Problem Posing

We shall illustrate our previous points by describing one of the ways in which an extranet can enhance students' (and teachers') mathematical learning within an international community. The domain we have chosen is data handling. Students in the eighth and ninth grades participated in collaborative activities that involved (1) data generation and exploration, and (2) statistical problem posing and critiquing.

Data Generation and Exploration (Inquiry)

The students generated their own data by designing a survey that was to be completed by each of their peers in all participating countries. Each student contributed to the design of the survey by posting suggested questions to an area within the Student Discussion Forum. The students found this "brainstorming" component both meaningful and motivating. As the teacher in Oxford (England) commented in our own Discussion Forum, "The [Web-based] brainstorming, I thought, was very successful.... I am sure that one of the things that contributed to the brainstorming being good was the fact that they [the students] knew it was for an *international* survey."

The final survey was created after all participants agreed on which particular questions should be included. This consensus was achieved through class discussions followed by forum discussions. We encouraged the students to consider survey questions that would generate different kinds of numerical data and that would yield interesting cross-cultural comparisons. The final survey questions included simple yes or no questions (e.g., "Do you believe in the death penalty for serious offences?" and "Are you in favor of the monarchy?"), nominal or "name" questions (e.g., "Who is your favorite Spice Girl?"), ordinal or "ranking" questions (e.g., "How concerned are you about the environment?: very concerned, concerned..."), and cardinal questions (requesting discrete values such as "How many children would you like to have of your own some day?" and continuous values such as "How long is your hair in centimeters from your scalp?"). The survey was then placed in a special zone on our extranet site and filled out, online, by every student.

A principal component of our extranet is its ability to process automatically data submitted by users to online forms. The raw data from the students' survey responses subsequently became available in a new zone and formed the basis of their statistical investigations and problem posing. The first step in these investigations was deciding how to analyze the raw data and whether such analysis might lead to important decisions and discoveries. To assist students in their decisions, we placed two simple questions on their Discussion Forum, to which each student, or pairs of students, responded:

1. In exploring the data, we noticed that..., and
2. We then wondered...

The students' responses to these questions stimulated interesting exchanges, especially on issues pertaining to cross-cultural differences. For example, a group of Australian students were surprised when the data appeared to suggest that the students in the U.K. were more concerned about the environment than they were. So, they wondered:

> Why would the U.K. students be more concerned about the environment than the Australians? In their responses, are they referring just to their country or to the whole world? Would the statistics and data change in the future? If so, how?

Problem Posing and Critiquing

What the students "wondered about" formed the basis of their statistical problem posing. In preparation for this, students and teachers in each class talked about mathematical problems in general. They then critiqued some sample problems. Some of the general issues discussed included these:

- How would you define a mathematical problem?
- What features does a problem require for it to be considered a mathematical problem?
- What do you consider makes ... (1) A good mathematical problem? (2) An appealing mathematical problem, that is, one that you would *want* to solve? (3) A challenging problem?
- When you construct your own problems from the issues you wondered about, what points will you need to consider?
- Are there additional points you need to think about when you are creating problems for your friends in other countries?
- What types of questions might you ask in your problems?
- What types of answers might you request in your problems? (Will there be more than one answer?)
- What are some ways in which you might ask your friends to display their solutions to your problems?

The students subsequently worked on posing their own problems, using survey data of their choice. Their problems were to require the solver to make a decision or discovery on the basis of the data referred to in the problem. We also encouraged the students to include "thought provoking" questions in their problems, that is, questions that required the solver to think beyond the data (e.g., "Do you think the percentage would differ if the survey were conducted in other countries?" "Why do you think the U.K. students spend less time on the computer than the Australian students?").

The students first shared their problem creations with their classmates and teachers for initial feedback prior to publishing them in the Problem Forum (this preliminary review also enabled teachers to ensure that appropriate conduct for Internet publishing was followed). The students gave attractive titles to their problems, which facilitated the identification of problems and also increased their motivation to tackle a problem. Some examples of the students' problem creations appear in table 7.2.

TABLE 7.2
Examples of Student-Generated Problems

Authors: Catherine, Megan, and Nicole, Grade 9—Brisbane, Australia
Title: Variety is the SPICE of life

It was a quiet day in England, when Mel C (sporty spice) awoke from her slumber and approached the Spice Girl's magic mirror. She questioned the mirror, "Mirror, mirror on the wall, who is the most popular Spice of all?" The mirror replied, "Look at the data and you will see, it's not Mel C, maybe Mel B!" Solve the following questions:

1) Who is the most popular Spice Girl to Australian girls?
2) Who is the most popular Spice Girl to Australian boys?
3) Who is the most popular Spice Girl to UK girls?
4) Who is the most popular Spice Girl to UK boys?
5) Who is the most popular Spice Girl overall?

Is Mel C right? Is the mirror right? Or are they both wrong?
NB It may help to put the first four questions in percentage form, in order to solve question 5.

———-

additional parts posted:
What formula's did you use in order to solve this problem?
Were you expecting this outcome? Why?

ACTUAL DATA

(Collected by online survey):

Favorite Spice Girl	Number of Australian Boys	Number of Australian Girls	Number of UK Boys	Number of UK Girls
Emma	6	10	4	6
Geri	0	2	12	1
Mel B	0	0	2	8
Mel C	0	1	1	1
Victoria	0	2	10	2
Don't Know	7	22	3	14
Total	13	37	32	32

Table 7.2 (cont.)

Author: Elizabeth—Toronto, Canada
Title: Dozing in Math

One Morning Mrs. [Teacher's Name] came into her classroom to see that some of her students were sleeping in her class! "Does the amount of sleep my students get on weeknights affect their alertness in my class?" Mrs [Teacher's Name] asked.

Answer the following questions:

a. Find the range of bedtimes on weeknights for Canadian and Australian results.
b. Find the median of bedtimes on weeknights for both countries.
c. Find the mean bedtime for each bedtime on the weeknights for both countries.
d. Find the average bedtime for each country and find out which country would be more alert for [Teacher's Name] math class.

ACTUAL DATA
(Collected by online survey):

Bed Time	Number of Students (Australia) weeknight	Number of Students (Canada) weeknight
9:00	12	1
9:30	19	5
10:00	5	6
10:30	10	7
11:00	2	2
11:30	1	0
12:00	0	3
12:30	1	1
1:00 am or later	0	0
Total	50	25

Each student in each of the participating classes then selected problems from the Problem Forum, accessing the necessary data to complete them. The students were very keen to try problems created by their peers in other countries ("It is good to see how kids in different countries have different ideas and values"). Interestingly, the students indicated that it was the *subject* of a problem, rather than the mathematics involved, that determined whether or not they found the problem a worthwhile and interesting one to solve. Students also liked problems that incorporated a thought-provoking question, as Megan explained: "It is good to be able to explain *why* in a non-mathematical term."

On solving a problem, the students completed an online critique (see fig. 7.2). These critiques were both highly motivating and important to the students' data-handling development. As indicated in figure 7.2, the solver offers constructive criticism on various aspects of the authors' problem, including its clarity and difficulty, and most important, offers suggestions on ways the

St. Rita's Authentic Data Critique Forum

Message #45
From: Katie
Problem: Mickey and Minney's dilemma
Date: 11/14/97, 12:37:20
Solution:
In Australia we counted the number of people that completed the chart was 50. Then we counted the number of people that agreed with the capital punishment. This was 37.
37*100/50 = 74%
Therefore 74% of the students who completed the chart in Australia were in favour of capital punishment.

In England 30 students out of 66 students who completed the chart said yes to capital punishment.
30*100/66 = 45.5%
Therefore 45.5% of all students who completed the chart in England were in favour of capital punishment.

I believe in every country you asked this survey in, you would get different results. Each country is brought up to believe in different values and beliefs. Every family and or person has their own opinion on the matter. My question is, Do Australian's believe in this method of dealing with criminals because this method is already in progress? Would there be similar results in the U.S.?

Overall, the problem is: vg

What I like most about the problem:
I liked the issue you chose, and how you could get very good ideas because most people have very strong opinions on this matter

What I liked least about the problem:
I think that this problem could have been a bit more complicated because it didn't take me very long.

Can the problem be solved? yes

The mathematics is too easy
What was difficult?
I didn't find much about this problem that was hard.

The wording is perfectly clear
What was confusing?
I didn't find much about the problem confusing.

Is the problem interesting? fairly clear
What was interesting?
As I have already said, I like the issues the are concerned in this questions.

Suggestions for improving/extending the problem:
They could have asked additional questions like, "What percent of those people are female/male?"

As I said above, you could have made the problem more difficult by adding additional questions.

Other comments:
Overall I thought this question was pretty good.

Reply/Next Message/Previous Message/Return to List of Messages

Fig. 7.2. A sample of a completed online critique

original problem could be improved and extended. The problem being critiqued in figure. 7.2, titled "Mickey and Minnie's Dilemma," is as follows:

> Mickey Mouse was wondering if the percentage of those in favor of the death penalty would be the same in the UK as in Australia. Minnie Mouse asked the question, "Would the percentage differ if the survey was conducted in different countries?" Can you answer these dilemmas for Mickey and Minnie?

Once students received their peers' critiques, they made improvements to their original problems, submitting the new version to replace the old on the Problem Forum.

Numerous cross-cultural discussions emerged from the students' problems, with several students extending these discussions beyond school time. Problems that drew mixed reactions from the students included those that addressed the survey data on the death penalty. Some students questioned the appropriateness of the death penalty as a topic for a problem: "It's not a very nice topic," "Mickey and Minnie should not be thinking about the death penalty." Other students were more supportive of the topic, such as Katie: "The statistics in the problem were of a valid topic and proved to be quite interesting…you could make it more interesting if you add other countries' statistics, which would make it more challenging as well." Other students made inappropriate statistical assumptions, such as Sarah and Ashley's comment: "We don't like the fact that we were unaware that if Australia has the death penalty or not. Because if they did, then of course the percentage would be much higher than the UK who does not have the death penalty."

Concluding Points

As educators, we need to ask ourselves how we can best make use of the Internet in fostering students' mathematics learning in this new century. In this article we have discussed just one of the many applications of this rapidly developing technology. Specially designed, semiprivate extranets present an excellent medium through which students and teachers around the globe can share data, exchange mathematical ideas, undertake collaborative investigative projects, publish a range of student-generated work, and develop an appreciation for one another's cultures.

In the twentieth century, students' mathematical activities were almost exclusively derived from teachers and textbooks—and the mathematical communities that students experienced were classroom-based and textbook driven. As we set forth in this new century, we need to reflect on the sentiments of Yerushalmy, Chazan, and Gordon (1993, p. 117):

There is something odd about the way we teach mathematics in our schools. We teach it as if we expect that our students will never have occasion to make new mathematics. We do not teach language that way. If we did, students would never be required to write an original piece of prose or poetry.

In the twenty-first century, we hope the situation changes and we see it become common for students and teachers regularly to create mathematical activities to share with their counterparts in other schools, wherever these schools might be. However, as Scardamalia (1997) noted, a hallmark of student engagement in educational networking is "the production of knowledge of value to others, not simply demonstrations of personal achievement." The production of this knowledge should not be devoid of worthwhile mathematical content.

The issue at stake in incorporating extranets within the curriculum is *not* less attention to core content or to activities that promote understanding (in other words, this technology will not become a replacement for the curriculum); rather, extranets provide a "whole new layer of activity over-and-above such content" (ibid.). The challenge is to capitalize on this new layer. Regardless of how sophisticated the technology becomes, it will not automatically create the learning communities we consider to be fundamental to students' mathematical growth, namely, carefully nurtured and defined communities of inquiry. These communities need to be established within actual classrooms before online communities are established. The power behind online communities rests with the participating teachers, whose communication on collaborative classroom activities and projects holds the key to exploiting extranet technology. Teachers who are motivated to challenge their students, to take their students' mathematical thinking and understanding beyond the confines of the class textbook, and to search actively for ways to enrich their own teaching of mathematics have this power.

REFERENCES

Bernard, Ryan. *The Corporate Intranet.*. 2nd ed. New York: John Wiley & Sons, 1998.

Brown, Ann L., and Joseph C. Campione. "Psychology Theory and the Design of Innovative Learning Environments: On Procedures, Principles, and Systems." In *Innovations in Learning,* edited by Leona Schauble and Robert Glaser, pp. 289–325. Mahwah, N.J.: Lawrence Erlbaum Associates, 1996.

Cudmore, Donald H., and Lyn D English. "Using Intranets to Foster Statistical Problem Posing and Critiquing in Secondary Mathematics Classrooms." Paper presented at the Annual Conference of the American Educational Research Association, San Diego, April 1998.

Lesh, Richard, M. Hoover, Bonnie Hole, Anthony Kelly, and Thomas Post. "Principles for Developing Thought-Revealing Activities for Students and Teachers." In *Research Design in Mathematics and Science Education,* edited by Richard Lesh, Barbara Lovitts, and Anthony Kelly. Mahwah, N.J.: Lawrence Erlbaum Associates, in press.

National Council of Teachers of Mathematics. *Curriculum and Evaluation Standards for School Mathematics.* Reston, Va.: National Council of Teachers of Mathematics, 1989.

———. *Principles and Standards for School Mathematics.* Reston, Va.: National Council of Teachers of Mathematics, 2000.

Pfaffenberger, Bryan. *Building a Strategic Extranet.* Foster City, Calif.: IDG Books, 1998.

Romberg, Thomas A., and Andrew Collins. "The Impact of Standards-Based Reform on Methods of Research in Schools." In *Research Design in Mathematics and Science Education,* edited by Richard Lesh, Barbara Lovitts, and Anthony Kelly. Mahwah, N.J.: Lawrence Erlbaum Associates, in press.

Scardamalia, Marlene. "Networked Communities Focused on Knowledge Advancement." Paper presented in the symposium "Collaboration, Communication, and Computers: What Do We Think We Know about Networks and Learning?" at the Annual Meeting of the American Educational Research Association, Chicago, 1997.

Silver, Edward A. "Moving beyond Learning Alone and in Silence: Observations from the QUASAR Project Concerning Communication in Mathematics Classrooms." In *Innovations in Learning,* edited by Leona Schauble and Robert Glaser, pp. 127–60. Mahwah, N.J.: Lawrence Erlbaum Associates, 1996.

Yerushalmy, Michael, Daniel Chazan, and Myles Gordon. "Posing Problems: One Aspect of Bringing Inquiry into Classrooms." In *The Geometric Supposer: What Is It a Case Of?,* edited by Judah Schwartz, Michael Yerushalmy, and Beth Wilson, pp. 117–42. Hillsdale, N.J.: Lawrence Erlbaum Associates, 1993.

8

As the Century Unfolds
A Perspective on Secondary School Mathematics Content

Johnny W. Lott

Terry A. Souhrada

THROUGHOUT the twentieth century the national debate about what constitutes good school mathematics was continuous. Changes in our society and the demographics of education as well as concerns about student achievement have spurred these discussions. Yet after many attempts at reform and returns to the status quo, the debate still rages. This article looks back at this century-old debate and argues that the current school mathematics curriculum does not serve well our students entering the twenty-first century.

At the 1902 annual meeting of the American Mathematical Society, E. H. Moore, the retiring president, gave an address entitled "On the Foundations of Mathematics" in which he recommended changes in the school mathematics program (Moore 1967, p. 366):

> I hold ... that by emphasizing steadily the practical sides of mathematics ... it would be possible to give very young students a great body of the essential notions of trigonometry, analytic geometry, and the calculus. This is accomplished, on the one hand, by the increase of attention and comprehension obtained by connecting the abstract mathematics with subjects which are naturally of interest to the [child], so that, for instance, all the results obtained by theoretic process are capable of check by laboratory process, and, on the other hand, a diminution of emphasis on the systematic and formal sides of instruction in mathematics.

Later in his address, Moore specifically expressed concern with the way that mathematics was taught in secondary school (p. 368):

> Engineers tell us that in the schools algebra is taught in one water-tight compartment, geometry in another, and physics in another, and that the student learns to

appreciate (if ever) only very late the absolutely close connection between these different subjects, and then, if he credits the fraternity of teachers with knowing the closeness of this relation, he blames them most heartily for their unaccountably stupid way of teaching him.

Going further, Moore called for teaching mathematics as a laboratory discipline and proposed the following pedagogy (p. 371):

> The instructor utilizes all the experience and insight of the whole body of students. He arranges it so that the students consider that they are studying the subject itself, and not the words, either printed or oral, of any authority on the subject. And in this study they should be in the closest cooperation with one another and with their instructor, who is in a desirable sense one of them and their leader.

In the language of today, Moore was calling for mathematics to be integrated and taught using an inquiry approach wherein learning would be a cooperative effort with the instructor acting as a fellow learner. These same messages were being echoed at the end of the twentieth century (American Mathematical Association of Two-Year Colleges 1995, pp. 4–5), giving the distinct impression that the same debate is cycling. The cycle needs to be examined through other lenses across the years to develop an understanding of the foundation on which to view the future.

OTHER CALLS FOR REFORM

Moore was not the only person calling for reform in the school mathematics curriculum early in the twentieth century. At the 1904 National Education Association meeting of the Department of Superintendents, Frank McMurry, in his presentation entitled "What Omissions Are Desirable in the Present Course of Study?" described the use that adults made of the mathematics being taught (National Council of Teachers of Mathematics [NCTM] 1970, p. 110). Whereas Moore looked at an integration of subject matter, McMurry suggested that some topics needed to be deleted from the mathematics curriculum. This debate too continued throughout the twentieth century, culminating with intense debate over the "decreased attention" recommendations contained in the *Curriculum and Evaluation Standards for School Mathematics* (NCTM 1989, p. 127) and the use of calculators in lieu of traditional computational algorithms.

In 1916 the National Committee on Mathematical Requirements was organized as a committee of the Mathematical Association of America (MAA) to help move forward reform in the teaching of mathematics. The committee wrote the following (MAA 1923, p. 395):

> The primary purposes of the teaching of mathematics should be to develop those powers of understanding and of analyzing relations of quantity and of space which are necessary to an insight into and control over our environment and to

an appreciation of the progress of civilization in its various aspects, and to develop those habits of thought and of action which will make these powers effective in the life of the individual.

The committee recognized that the teaching of mathematics had to serve three purposes: (1) practical, (2) disciplinary, and (3) cultural. From the beginning, the committee wrote that the three purposes were not distinct and that "any truly disciplinary aim is practical and, in a broad sense, the same is true of cultural aims" (p. 391). As a result, the committee saw advantages in developing integrated mathematics courses and wrote (p. 397):

> The newer method of organization enables the pupil to gain a broad view of the whole field of elementary mathematics early in his high-school course. In view of the very large number of pupils who drop out of school at the end of the eighth or the ninth school year or who for other reasons then cease their study of mathematics, this fact offers a weighty advantage over the older type of organization under which the pupils studied algebra alone during the ninth school year, to the complete exclusion of all contact with geometry.

The committee also recognized a desire of teachers to have a specific syllabus in place that would tell them what topics to teach in what order with specific time allotments for each. However, the report states, "This desire can not be met at the present time for the simple reason that no one knows what is the best order of topics, nor how much time should be devoted to each in an ideal course" (p. 401). This desire and same discussion continued throughout the twentieth century.

The push for reform continued into the 1940s and 1950s. In 1945, "The Second Report of the Commission on Post-War Plans: The Improvement of Mathematics in Grades 1 to 14" was published (NCTM 1945). It stated, "The high school needs to come to grips with its dual responsibility, (1) to provide sound mathematical training for our future leaders of science, mathematics, and other learned fields, and (2) to insure mathematical competence for the ordinary affairs of life to the extent that this can be done for all citizens as a part of a general education appropriate for the major fraction of the high school population" (p. 195).

Though some things were being done well in 1945, on 17 October 1946, Presidential Executive Order 9791 called for a report on, and recommendations for, the status of the national supply of trained personnel for programs of scientific research and development. In response to the order, appendix 2 in volume 4 of the report *Science and Public Policy* (Steelman 1947) was devoted to and titled "The Present Effectiveness of Our Schools in the Training of Scientists." This appendix not only addressed the supply and training of scientists but looked at the mass education of citizens. The conclusion was that "our present efforts are far too weak and apparently not sufficiently introspective to produce an awareness of the basic nature of science and of its importance in the general social enterprise" (p. 57). Among the findings were the following:

1. The teaching of mathematics and science in elementary school—although following textbooks—often follows them without any integrated plan extending from grade to grade in schools.
2. Grade placement of topics is by tradition and teacher preference rather than any systematic selection.
3. High school graduates may have little or no education in the physical sciences, and even those who go to college and graduate may remain scientifically (and mathematically) illiterate.
4. Potential scientists are lost because of a lack of encouragement and guidance.

A summary of studies, such as that of Schorling in 1931, was used to back up the findings in this report (Steelman 1947, p. 63):

> Of 2,693 freshmen at the University of Michigan, only 66 percent were able to multiply 2 1/2 by 3 1/4; only 68 percent were able to reduce 3/20 to a decimal fraction; and only 42 percent could factor $6x^2 + 7x + 2$; the best showing—81 percent—was made on dividing 7,642.38 by 1,000!

Among the recommendations for elementary school in the report was a call for research on when and how to teach concepts, principles, and skills in mathematics and science; desirable facilities; materials and aids to use; and procedures for integrating concepts from various sciences, mathematics, and other areas. At the high school level, there was a call for the to-be-created National Science Foundation to conduct a study to determine the mathematical needs of students with scientific aptitudes and to design a course of study for them. In addition, the general recommendations called for a study of the entire secondary school curriculum with regard to the general education needs of all students (Steelman 1947).

The subsequent establishment of the National Science Foundation in 1950, the creation of the University of Illinois Committee on School Mathematics in 1951, the work of the School Mathematics Study Group in 1958, and the 1959 work of the Commission on Mathematics appointed by the College Entrance Examination Board (CEEB) moved the school mathematics programs in this country in a direction that persisted until late in the century. Most school mathematics programs developed after the 1950s contained mathematical topics evolved from the suggested course of study in the *Program for College Preparatory Mathematics* in 1959 (CEEB). This course of study was then developed by the School Mathematics Study Group (SMSG 1961) and was later influenced by the report *Goals for School Mathematics* (Cambridge Conference on School Mathematics 1963), the National Advisory Committee on Mathematical Education (NACOME)

1975 report *Overview and Analysis of School Mathematics: Grades K–12,* and the Basic Mathematics Skills and Learning Conference held in Euclid, Ohio, in 1975. Included in this evolving curriculum framework was an increased focus on the theory underlying school mathematics, with the apparent assumption that by including theory, both skills and understanding would greatly increase. Topics and content areas were unified under the familiar course titles of Algebra 1, Geometry, Algebra 2, and Precalculus in the 1960s through the 1980s. However, debates about content in school mathematics continued throughout this period, causing many changes. Although these debates resulted in much of the focus of the 1960s on teaching mathematical theory being severely limited by the 1980s, many other aspects of the framework, such as course structure and delivery, remained in schools (Usiskin 1985).

Almost a century after Moore's and McMurry's presentations, the school mathematics debate continued. In 1989, the National Council of Teachers of Mathematics attempted to reform mathematics education with the publication of the *Curriculum and Evaluation Standards for School Mathematics* (NCTM 1989). The document included a reevaluation of the current content of study. Additionally, the *Standards* proposed an improvement in approaches to the instruction and assessment of the content. Companion pieces to the 1989 *Standards* were the *Professional Teaching Standards for School Mathematics* (NCTM 1991) and the *Assessment Standards for School Mathematics* (NCTM 1995). Additionally, in its 1989 publication *Everybody Counts: A Report to the Nation on the Future of Mathematics Education,* the National Research Council (NRC) reported that "mathematics education, in contrast [to mathematics and science per se], has been constrained by societal forces to such a degree that it has hardly changed at all. This contrast in the pace of change virtually ensures that mathematics education is perpetually out of date" (NRC 1989, p. 76). The release of NCTM's *Standards* documents along with the results of the Second International Mathematics Study (Flanders 1994), which showed that students in the United States did not compare favorably in mathematical ability with students from around the world, further fueled the debate and efforts to reform.

The twentieth century culminated in a series of National Science Foundation–sponsored curriculum development projects surrounded in controversy over suggested reforms and with the National Council of Teachers of Mathematics revising its earlier *Standards* with the *Principles and Standards for School Mathematics: Discussion Draft* (NCTM 1998). The twentieth century had seen high school mathematics change from being primarily arithmetic not required by all to a polyglot of topics in algebra and geometry required for all as noted in *Learning from TIMSS: Results of the Third International Mathematics and Science Study, Summary of a Symposium* (Beatty 1997).

Other Considerations for School Mathematics Reform

With a renewed demand for reform, it is time once again to think about needed changes for the twenty-first century. School mathematics can be examined through a series of lenses for differing needs driving reform: collegiate mathematics, needs of the workplace, a review of the school mathematics curriculum, and an assessment of the curriculum.

Collegiate Mathematics

To begin completing a picture of the needs of mathematics education, it is instructive to consider what mathematics college students were taking at the end of the 1990s and what their perceived needs were. In *Statistical Abstract of Undergraduate Programs in the Mathematical Sciences in the United States: Fall 1995 CBMS Survey* (Loftsgaarden, Rung, and Watkins 1997), it is reported that about 57 percent of students in undergraduate mathematics in two-year colleges, four-year colleges, or universities take either remedial or precalculus mathematics. This is important in the complete mathematics education picture because this level of mathematics is typical of high school courses taken during the junior year. Also of note is the fact that 46 percent of the total collegiate mathematics enrollment is at two-year colleges. What is seen in this study is that though much of the secondary school coursework is geared toward preparing the college-intending student to take calculus, much of the mathematics being taught to college students is secondary school material and hence relatively few college students ever take a calculus course. This fact alone should make reform in the mathematics curriculum of prime importance.

Needs of the Workplace

Factors involving the workplace are equally as important in understanding the state of mathematics as factors involving academia. The workplace has evolved since Moore's and McMurry's calls for change. In the latter part of the twentieth century, the traditional mathematics curriculum required that industry spend as much on remedial mathematics education for its workers as that spent on mathematics education in schools, colleges, and universities (NRC 1989). People without a good understanding of mathematical concepts are not becoming productive members of society (Kearns 1995). Technology has become a mainstay in businesses and industries, with more employees directly involved in the problem-solving process (Katz and Jury 1997). With these changes, mathematics educators must answer the following questions to prepare students: How do the changes in the workplace and education's responsibility translate into the mathematics classroom? Will the current traditional mathematics curriculum satisfy this new responsibility? As Kearns points out, "No one should be left behind. No one should be denied becoming a part of the workforce of tomorrow and the higher quality of life that should go with it" (1995, p. 326).

According to Ross (1995) and others, the traditional school mathematics program fails to prepare many if not most of our students for a workplace where "mathematics has ... become the language of the technical workforce" (Steen and Forman 1995, p. 224). In addition, Wallhaus (1996) suggests that the workplace skills and the traditional practices in mathematics education are opposed to each other: the integration of skills in the workforce versus the independence of disciplines in academia; cooperation and teamwork versus competitiveness and individualism; and tolerance for ambiguity versus precision and stability.

Of such concern is education in the nation's schools that the Business Roundtable (1990) adopted nine essential components for a new school system in its public policy agenda. One assumption of the first component follows (p. 4):

> Curriculum content must lead to higher order skills, and instructional strategies must be those that work. What children learn should be commonly challenging. We must focus them on thinking, problem solving, and integration of knowledge. ...When we fail with a single child or a class or school, we must recognize we do not yet have the proper mix of how, where, when and who.

To make the transition from school to work, school mathematics should resemble work mathematics and be developed within, and grow from, meaningful contexts (Steen and Forman 1995; Katz and Jury 1997).

High-technology, multiethnic, rapidly shifting markets no longer call for a static set of "basic" skills. More important are schools that nurture multiple perspectives, multiple representations, the decision-making process, and drawing on one's own experiences (Katz and Jury 1997). The skills necessary for the next century must go beyond arithmetic, algebra, and geometry and include the practice and use of basic workplace skills that involve leadership, reliability, interpersonal communications, teamwork, and the ability to function in a rapidly changing environment (Wallhaus 1996). Students must also learn to make appropriate use of all reasonably available tools for learning mathematics. Technology removes the drudgery of computation and gives people the ability to solve difficult, challenging, and meaningful mathematics problems that require a deep understanding (Seeley 1995).

The School Mathematics Curriculum

The consideration of what mathematics students take in colleges and the needs of the workplace at the end of the century raises questions about the school curriculum. If as indicated above, the mathematics curriculum is not meeting the needs of our students, we should consider what mathematics the students are taking and the content of that mathematical curriculum. The data in table 8.1, adapted from the *NAEP 1996 Trends in Academic Progress* (Campbell, Voekk, and Donahue 1997, p. 85), show the percent of seventeen-year-old students taking high school mathematics. If there are problems with students' preparation, the majority of the problems lie with the courses listed in table 8.1 and primarily with the algebra 1, geometry, and algebra 2 courses.

TABLE 8.1
Seventeen-Year-Old Students' Highest-Level Mathematics Courses in 1996

Courses	Percent of Students
Prealgebra or General Mathematics	8
Algebra 1	12
Geometry	16
Algebra 2	50
Precalculus or Calculus	13

It is instructive to consider the actual content of the studied curriculum. In an analysis of TIMSS (Beatty 1997, p. 9), Schmidt made the following observation: "Mathematics curricula in the United States consistently cover far more topics than is typical in other countries." Schmidt also noted that the textbooks' topics seem to be jumbled instead of forming a coherent whole (Beatty 1997, p. 10). Additionally, on the basis of his review of middle school textbooks and an evaluation of the Second International Mathematics Study (SIMS), Flanders concluded that teachers primarily rely on textbooks as the sole source for instruction, that little is taught that is not in textbooks, and that authors and publishers should be encouraged to develop textbooks that present a more balanced access to the topics of the SIMS test (Flanders 1988, 1994). To see where problems exist in a broad-brush picture of the mathematics curriculum, it is important to examine the topics found in algebra 1, geometry, and algebra 2.

Prior to the 1980s, algebra 1 and algebra 2 texts were quite similar in both content and format (Corcoran et al. 1981; Gaughan et al. 1984). The mathematics topics included were designed to prepare students for calculus (CEEB 1959). As such, algebra 1 included an introduction to algebra involving variables, expressions and equations, linear and quadratic equations, polynomial expressions, factoring, and real numbers and their operations. Algebra 2 extended these topics in a similar fashion and included conic sections, logarithmic and exponential functions, sequences and series, complex numbers, and trigonometry. Additionally, some texts included the introduction of basic notions of probability and statistics.

In both algebra 1 and algebra 2 books, problem sets were often concluded with a few contrived "word problems" representing real applications of the mathematics introduced. After the introduction of the NCTM *Standards* documents, some algebra texts began to change. The number of practice problems was reduced, word problems were improved, and some sections were added to make minimal use of available technology. However, the topics and their presentations essentially remained unchanged.

Before the 1980s, geometry typically focused on definitions, axioms, and reasoning, with students immersed in the notion of proof. The proofs generally centered on congruent triangles, circles, parallel lines, and similarity. Sections of the geometry texts also focused on constructions and loci, perimeter, and area. Algebra, with the exception of a short introduction to coordinate geometry, was not evident in the course (Cummins et al. 1988). By the 1970s, transformational geometry had also begun to appear in secondary school texts (Coxford and Usiskin 1971), and alternative texts began to appear in which proof was de-emphasized and an exploratory approach to discovering geometric properties was implemented (Hoffer 1979; Serra 1987).

Other general observations about selected texts in the latter part of the twentieth century include the following: Technology was sometimes alluded to, computer and calculator sections were added to the text, and frequently there was an appendix or a section in a teacher's manual that dealt with the use of computers or some type of graphing calculator (Nichols et al. 1986).

An Assessment of the Mathematics Curriculum

A report from the National Center for Education Statistics released in 1997 indicates that twelfth graders performed better in 1996 than in the two previous National Assessment of Educational Progress (NAEP) testing cycles. The percents of students who scored at or above the basic level were 42 percent in 1990, 64 percent in 1992, and 69 percent in 1996 (Reese et al. 1997, p. 47). On the surface this appears to be strong evidence for maintaining the status quo in school mathematics, but these figures deserve closer examination.

In 1991, the National Assessment Governing Board (NAGB) of NAEP adopted definitions for three achievement levels—Basic, Proficient, and Advanced—on which to base its biannual reports. These levels were defined for grades 4, 8, and 12, with the Proficient level being the national goal for student achievement (NAGB 1991).

At a time when 63 percent of seventeen-year-old students are taking the equivalent of algebra 2 or above (Campbell et al. 1997) and the number of students at or above the Basic level was increasing, still 84 percent of all twelfth-grade students tested scored below the Proficient level on the 1996 NAEP exam, with 80 percent of the white students, 96 percent of the black students, 94 percent of the Hispanic students, and 97 percent of the Native American students falling below this level (Reese et al. 1997, p. 55). One reason for this failure may very well have to do with the materials and methodology used in U.S. classrooms.

Olson (1999, p. 1) reported that a 1997 video study conducted in connection with TIMSS found the following:

> Everywhere it seemed, American teachers taught using ... the same uninspiring methods. Teachers largely drilled their students on low-level procedures. In not one of the 81 videotaped classes did students perform a mathematical proof...while most of the videotaped teachers claimed to be using practices recommended by the National Council of Teacher of Mathematics ... the analysts saw almost no evidence of this.

Teachers and textbooks in the 1990s were still using the same instructional strategies that had been used since the 1950s and before. Classes still reviewed previous material followed by the teacher stating the concept of the day, students did individual seat work that imitated the teacher's examples, and the teachers gave homework consisting of more of the same type of drill problems assigned for the next day (Olson 1999).

This methodology is in conflict with studies conducted in the 1980s and before (Resnick and Ford 1981; Suydam and Dessart 1980; Farrell and Farmer 1988) that indicated that the practices of presenting mathematical concepts through stand-alone examples and repetitious practice sets do little to promote the understanding of these concepts. These studies found that drill did little to affect learning and the transfer of learning to other areas (Resnick and Ford 1981).

OTHER CONSIDERATIONS OF MATHEMATICS EDUCATION AT THE TURN OF THE CENTURY

After examining curricular materials and seeing that the courses based on those materials are indeed what is taken by students, we raise a natural question: "How did the materials and courses become such a jumbled collection of topics?" This question deserves serious research. Anecdotally, from conversations with commercial mathematics textbook editors, it appears that textbooks and exercises are designed to meet the guidelines of the state textbook adoption requirements of such larger states as California, Texas, and Florida. The state guidelines attempt to do what the MAA noted that teachers wanted in 1923: specific topics to be taught in a specific order for a specified amount of time. *The Western Canadian Protocol for Collaboration in Basic Education* (The Crown 1996) is a Canadian attempt to do this.

If we consider the desires of teachers for specificity in curricular guidelines and the needs of collegiate mathematics both juxtaposed with the ever-changing needs of the workplace, it is a wonder that a fixed state curriculum could ever be adopted. However, state textbook adoption guidelines are often quite specific about what texts must contain. Some of the state adoption guidelines are so specific that there is little room for change or the variability of topics in the curriculum. For example, the Texas adoption guidelines are defined to the extent that if a textbook does not meet 100 percent of the guidelines, then the text can be only provisionally accepted for being on the state-supported text list for schools (Texas Education Agency 1995).

The impact of national standards on the content and coherence of textbooks still remains to be seen. The NCTM *Standards* documents (1989, 1991, 1995) provided overall guidance for school mathematics in the areas of curriculum, professional teaching, and assessment, and many states, such as Texas and Virginia, adopted state standards that reflected them. By 1998, however, states such as California began to have a backlash against the less-restrictive NCTM *Standards* documents, and school boards began demanding a swing back to a more traditional curriculum (Wu forthcoming).

PERCEIVED NEEDS AND NEXT STEPS FOR ACADEMIC CHANGE

As early as 1985, Usiskin pointed out that in general, students were not taking enough mathematics in high school and that even students who did take three years of college-preparatory mathematics did not learn the uses of that mathematics. As a result of these observations, Usiskin asked for a revolution in school mathematics (1985). Knowing the needs of academia, the workplace, and the reality of student achievement, we can see that the need for change is still true today.

To reexamine and update its three *Standards* documents, the NCTM released in 1998 *Principles and Standards for School Mathematics: Discussion Draft* for public comment. This document is being prepared to move the mathematics community another step in the direction of reform. Recognizing some of the problems mentioned by Usiskin and realizing that technology is changing the face of mathematics education at the end of the twentieth century, NCTM invited some of its best thinkers to suggest *Standards* revisions to prepare students and teachers for the twenty-first century (Howe

1998a, 1998b). The *Principles and Standards* draft called for a focus on the following areas: number and operations; pattern, functions, and algebra; geometry and spatial sense; measurement; data analysis, statistics, and probability; problem solving; reasoning and proof; communication; mathematical connections; and representation. These areas are spread across the pre-K–2, 3–5, 6-8, and 9–12 grade bands. Prior to the release of the *Principles and Standards* draft, the National Research Council, in conjunction with NCTM, the National Science Foundation, and the Mathematical Sciences Education Board, sponsored an algebra symposium in 1997 and a middle-grades mathematics conference in 1998 to consider needed reforms.

CONCLUSION

Evidence from the workplace, colleges and universities, and test results illustrates that the traditional mathematics curriculum is not serving today's students and will not serve students as the twenty-first century progresses. This curriculum is not producing the problem solvers that business and industry need. Topics taught as isolated bits of information devoid of real-world applications and taught in isolation from other concepts in the same discipline still characterize the mathematics delivery. A survey of the twentieth-century debate reveals that although change has taken place in the content and demographics of the curriculum, many if not most of the same concerns appear to recur.

As we enter the twenty-first century, the need for change is as strong as ever. Proposed changes include a revamping of most of what mathematics teachers teach, the style in which they teach, and the tools that they use while maintaining a rigorous mathematics curriculum. Requisite with such change is a revamping of the testing and assessment that should be used. Finally, such change demands that colleges and universities rethink their entire freshman mathematics curriculum. At all levels we must answer the following questions: What topics will make up the mathematics curricula of the twenty-first century? How will these topics be taught? How will current teachers of mathematics be properly prepared to teach the school mathematics of the new century? What changes must be implemented in the preparation of future mathematics teachers? What role in mathematics education will technology play? How will students' learning be assessed?

These questions, posed as they are at the end of a century of debate, indicate that an era has arrived in which a new *Commission on College Preparatory Mathematics* report is required. This era will be characterized by college-preparatory mathematics influencing changes in the college program.

To make those changes, the advice given by Moore and McMurry at the beginning of the twentieth century and others throughout the century is equally as valid now as it was then:

1. Integrate the mathematical concepts so that discrete courses of algebra, geometry, trigonometry, statistics, and so on, do not exist.
2. Determine the mathematical topics necessary for the new century and delete those topics that are no longer integral to the curriculum.
3. Use the tools of the day, including technology, to enhance the learning of mathematics.
4. Teach mathematics with a variety of techniques to meet the needs of all students.
5. Adopt new ways of assessing students' learning and methods of teaching.
6. Accept the reality that the mathematics curriculum is not fixed and will change frequently.
7. Ensure that a continuous, systematic effort is made to keep the public informed about what is needed in mathematics and that mathematical illiteracy at any level should not be tolerated.
8. Change teacher-preparation programs to meet the demands of new curricula.

Should we heed this advice? To ignore it will only guarantee that the debates of the twentieth century will persist into the twenty-second century.

REFERENCES

American Mathematical Association of Two-Year Colleges. *Crossroads in Mathematics: Standards for Introductory College Mathematics before Calculus.* Memphis, Tenn.: American Mathematical Association of Two-Year Colleges, 1995.

Beatty, Alexandra, ed. *Learning from TIMSS: Results of the Third International Mathematics and Science Study: Summary of a Symposium.* Washington, D.C.: National Academy Press, 1997.

Business Roundtable. *Essential Components of a Successful Education System: The Business Roundtable Education Public Policy Agenda.* New York: National Alliance of Business, 1990.

Cambridge Conference on School Mathematics. *Goals for School Mathematics.* Boston: Houghton Mifflin Co., 1963.

Campbell, J. K., K. R. Voekk, and P. L. Donahue. *NAEP 1996 Trends in Academic Progress.* Washington, D.C.: National Center for Education Statistics, 1997.

College Entrance Examination Board, Commission on Mathematics. *Program for College Preparatory Mathematics.* New York: College Entrance Examination Board, 1959.

Corcoran, Clyde L., Edward D. Gaughan, Norman E. Ladd, and Judith Salem. *Scott, Foresman Algebra First Course.* Glenview, Ill.: Scott, Foresman & Co., 1981.

Coxford, Arthur, and Zalman Usiskin. *Geometry: A Transformation Approach.* River Forest, Ill.: Doubleday & Co., Laidlaw Brothers, 1971.

The Crown in Right of the Governments of Manitoba, Saskatchewan, British Columbia, Yukon Territory, Northwest Territories, and Alberta. *The Common Curriculum Framework for K–12 Mathematics: Western Canadian Protocol for Collaboration in Basic Education.* Calgary, Alta.: Alberta Education, 1996.

Cummins, Jerry J., Margaret Kenney, Timothy D. Kanold, Alice R. Kidd, and James L. Smith. *Merrill Informal Geometry.* Columbus, Ohio: Merrill Publishing Co., 1988.

Farrell, Margaret A., and Walter A. Farmer. *Secondary Mathematics Instruction: An Integrated Approach.* Providence, R.I.: Janson Publications, 1988.

Flanders, James R. "How Much of the Content in Mathematics Textbooks Is New?" *Arithmetic Teacher* 35 (September 1988): 18–23.

———. "Textbooks, Teachers, and the SIMS Test." *Journal for Research in Mathematics Education* 25 (May 1994): 260–78.

Gaughan, Edward D., Clyde L. Corcoran, John F. Devlin, Harold D. Taylor, and Loretta M. Taylor. *Scott, Foresman Algebra Second Course.* Glenview, Ill.: Scott, Foresman & Co., 1984.

Hoffer, Alan R. *Geometry.* Menlo Park, Calif.: Addison-Wesley Publishing Co., 1979.

Howe, Roger. "The AMS and Mathematics Education: The Revision of the 'NCTM Standards.'" *Notices of the American Mathematical Society* 45 (February 1998a): 243–47.

———. "Reports of AMS Association Resource Group." *Notices of the American Mathematical Society* 45 (February 1998b): 270–76.

Katz, Mira, and Mark Jury. "Literacies in a Changing Workplace: A Look at the Uses of Literacy in a Multi-ethnic, High-Tech Electronics Factory." Paper presented at the annual meeting of the American Educational Research Association, Chicago, March 1997.

Kearns, David. "The Business and Industry Perspective." In *Seventy-five Years of Progress: Prospects for School Mathematics,* edited by Iris M. Carl, pp. 322–28. Reston, Va.: National Council of Teachers of Mathematics, 1995.

Loftsgaarden, Don, Donald Rung, and Ann Watkins. *Statistical Abstract of Undergraduate Programs in the Mathematical Sciences in the United States: Fall 1995 CBMS Survey.* Washington, D.C.: Mathematical Association of America, 1997.

Mathematical Association of America. *The Reorganization of Mathematics in Secondary Education: A Report by the National Committee on Mathematical Requirements under the Auspices of the Mathematical Association of America, Inc.* Washington, D.C.: Mathematical Association of America, 1923.

Moore, Eliakim Hastings. "On the Foundations of Mathematics." *Mathematics Teacher* 60 (April 1967): 360–74. (A reprinting of the 1902 address, first published in *Science,* 1903, and later included in *A General Survey of Progress in the Last Twenty-five Years,* First Yearbook of the National Council of Teachers of Mathematics, 1926.)

National Advisory Committee on Mathematics Education. *Overview and Analysis of School Mathematics: Grades K–12.* Reston, Va.: National Council of Teachers of Mathematics, 1975. (Available through the ERIC system, document number ED 115 512.)

National Assessment Governing Board. "NAGB Sets Standards for the 1990 NAEP Mathematics Assessment." *National Assessment Governing Board Bulletin.* Washington, D.C.: National Assessment Governing Board, June 1991.

National Council of Teachers of Mathematics. *Assessment Standards for School Mathematics.* Reston, Va.: National Council of Teachers of Mathematics, 1995.

———. *Curriculum and Evaluation Standards for School Mathematics.* Reston, Va.: National Council of Teachers of Mathematics, 1989.

———. *A History of Mathematics Education in the United States and Canada.* Thirty-second Yearbook, edited by Phillip S. Jones and Arthur F. Coxford. Washington, D.C.: National Council of Teachers of Mathematics, 1970.

———. *Principles and Standards for School Mathematics: Discussion Draft.* Reston, Va.: National Council of Teachers of Mathematics, 1998.

———. *Professional Teaching Standards for School Mathematics.* Reston, Va.: National Council of Teachers of Mathematics, 1991.

———. "The Second Report of the Commission on Post-War Plans: The Improvement of Mathematics in Grades 1 to 14." *Mathematics Teacher* 38 (May 1945): 195–221.

National Research Council. *Everybody Counts: A Report to the Nation on the Future of Mathematics Education.* Washington, D.C.: National Academy Press, 1989.

Nichols, Eugene D., Mervine L. Edwards, E. Henry Garland, Sylvia A. Hoffman, Albert Mamary, and William F. Palmer. *Geometry.* Austin, Tex.: Holt, Rinehart & Winston, 1986.

Olson, Steve. "Candid Camera." *Teacher Magazine* 10 (May/June 1999): 28–32. (Available online at www.edweek.org/tm/current/o8candid.h10.)

Reese, Clyde M., Karen E. Miller, John Mazzeo, and John A. Dossey. *NAEP 1996 Mathematics Report Card for the Nation and the States.* Washington, D.C.: National Center for Education Statistics, 1997.

Resnick, Lauren B., and Wendy W. Ford. *The Psychology of Mathematics for Instruction.* Hillsdale, N.J.: Lawrence Erlbaum Associates, 1981.

Ross, Ken. "President's Column: Back to Math Education Reform." *Focus* 15 (April 1995): 9–10.

School Mathematics Study Group (SMSG). *Report of a Conference on Future Responsibilities for School Mathematics.* Stanford, Calif.: School Mathematics Study Group, 1961.

Seeley, Cathy. "Changing the Mathematics We Teach." In *Seventy-five Years of Progress: Prospects for School Mathematics,* edited by Iris M. Carl, pp. 242–60. Reston, Va.: National Council of Teachers of Mathematics, 1995.

Serra, Michael. *Discovering Geometry: An Inductive Approach.* Berkeley, Calif.: Key Curriculum Press, 1987.

Steelman, John R. *Manpower for Research.* Vol. 4 of *Science and Public Policy, a Report to the President.* Washington, D.C.: U.S. Government Printing Office, 1947.

Steen, Lynn Arthur, and Susan L. Forman. "Mathematics for Work and Life." In *Seventy-five Years of Progress: Prospects for School Mathematics,* edited by Iris M. Carl, pp. 219–41. Reston, Va.: National Council of Teachers of Mathematics, 1995.

Suydam, Marilyn N., and Donald J. Dessart. "Skill Learning." In *Research in Mathematics Education,* edited by Richard J. Shumway, pp. 207–43. Reston, Va.: National Council of Teachers of Mathematics, 1980.

Texas Education Agency. "1995 Proclamation of the State Board of Education Advertising for Bids on Instructional Materials." Austin, Tex.: Texas Education Agency, 1995.

Usiskin, Zalman. "We Need Another Revolution in Secondary School Mathematics." In *The Secondary School Mathematics Curriculum,* 1985 Yearbook of the National Council of Teachers of Mathematics, edited by Christian R. Hirsch, pp. 1–21. Reston, Va.: National Council of Teachers of Mathematics, 1985.

Wallhaus, Robert. "The Roles of Postsecondary Education in Workforce Development: Challenges for State Policy." Paper prepared for the Wingspread Symposium, Racine, Wis., 22–24 February 1996. (ERIC Document Reproduction no. ED 394403)

Wu, Hung-Hsi. "The 1997 Mathematics Standards War in California." In *What Is at Stake in the K–12 Standards Wars: A Primer for Educational Policy Makers,* edited by Sandra Stotsky. Cambridge, Mass.: Peter Lang Publishing, forthcoming.

9

The Impact of *Standards*-Based Instructional Materials in Mathematics in the Classroom

Eric E. Robinson

Margaret F. Robinson

John C. Maceli

THE vision of school mathematics suggested by the National Council of Teachers of Mathematics (NCTM) *Standards* documents (NCTM 1989, 1991, 1995) has been interpreted and implemented in a variety of forms. Indeed, this is as intended. The *Standards* did not detail a lock-step approach to mathematics curriculum but rather recommended a general philosophy, direction, and challenge to improve the school mathematics education of our students. Nonetheless, the *Standards* suggested profound change in almost every aspect of the teaching and learning of mathematics. The *Principles and Standards for School Mathematics* (NCTM 1998) refines and elaborates on the messages detailed in the original documents, keeping the basic vision intact. This means that a true *Standards*-based mathematics education requires a considerable, consequential shift in what teachers do in the classroom and possibly a paradigm shift in their view of mathematics itself. Therefore, current and prospective teachers need to know not only what changes in mathematics education are contained in *Standards*-based instructional materials but also why these changes are present and how they affect students' learning and understanding.

The preparation of this article was supported, in part, by the National Science Foundation Grant No. ESI-9634085. The opinions expressed are those of the authors and not the Foundation.

In the 1990s, the National Science Foundation (NSF) funded a collection of large-scale, multiyear development projects of instructional materials at each of the elementary, middle, and high school levels (see appendix). Funding for these projects over a number of years allowed developers the time and resources to think carefully and deeply about the intent of the *Standards*. As a result, a new generation of curricula emerged that aligns with the *Standards* and implements their ideas. These curricula provide a variety of instructional models accessible to all students. Each curriculum has been through a rigorous process of design, pilot testing, field testing, evaluation, and redesign before reaching the commercial market.

These NSF-funded curricula serve as an excellent foundation for examining the impact of *Standards*-based instructional materials in the school mathematics classroom. These materials represent a profound departure from traditional textbooks of the past, both in content and in format. They cannot be adequately judged or reviewed by a cursory scanning. So in this article our intent will be to explicate several features of these materials, taken as a group, and their impact in the classroom on both teachers and students. Specifically, we will examine aspects of their philosophical focus, their instructional and pedagogical strategies, their approaches to algorithms, their choice of mathematical content, and their use of technology. Many of our conclusions are not new in the sense of novel ideas in themselves. What is significant is that these ideas are implemented consistently and cohesively throughout multiyear curricula.

To provide some experiential basis for our discussion, please spend a few minutes doing the activity below, which accompanies figure 9.1. It is encountered by students in the second year of a high school curriculum (Mathematics: Modeling Our World, Course 2, pp. 6–7). Because this activity was designed for a small group, it would be preferable for you to work as part of such a group, if possible.

ACTIVITY 1: THE NEW FIRE STATION

Welcome to Gridville! This small village has grown in the past year. The people of Gridville have agreed they now need to build a fire station. What is the best location for the fire station?

All towns and cities are different. Gridville is a simplified version of a small village with streets running parallel and perpendicular to one another. It represents the start of the modeling process. What you learn in Gridville may be useful in other settings that require finding a best location. The streets of Gridville are two-way streets represented by the lines of the grid. Houses are represented by points and are located at the intersections of grid lines in order to identify their locations easily. Fire trucks may only travel along grid lines. Diagonal movement is not allowed. It is simpler than a real town in its size and in its layout (geometry), but it still has houses and roads, and it needs a fire station.

THE TASK:

The map of Gridville shows the location of all houses. Determine the best location for Gridville's fire station, and write a persuasive argument defending your choice. The city leaders will follow the advice of the group that delivers the most convincing argument.

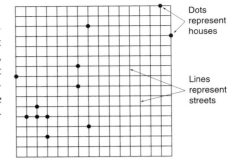

Fig. 9.1. Gridville

GUIDELINES:

1. Begin your argument by answering the questions, "What are important factors to consider in deciding the best location?" and "What does 'best' mean?"

2. You are encouraged to use charts, diagrams, tables, graphs, equations, calculations, and logical reasoning in making your decision.

3. Clearly state your choice of best location. Your summary should include the arguments and mathematics that support your decision. The summary should also explain how your charts, diagrams, tables, graphs, equations, calculations, and logical reasoning relate to the factors you considered and led your group to your choice.

4. Your written presentation may be posted in a display area in the classroom.

5. In addition to the written presentation, your group will give an oral presentation of approximately two minutes. The oral presentation should summarize your arguments and explain the reasons for your decision.

(Group presentations)

CONSIDER:

Answer the following questions based on the classroom presentations for Activity 1, *The New Fire Station*.

1. Which criteria were used most often to determine the location of the fire station?

2. Which factors invite mathematical investigation?

3. Even a model such as Gridville can be simplified further in order to study the essential elements of the location problem in detail. What are some ways you can simplify the Gridville model to investigate distance relationships?

(From COMAP Course 2, Modeling Our World, Student Textbook, 1st edition, by John Maceli, ©1998. Reprinted with permission of South-Western Educational Publishing, a division of International Thomson Publishing. Fax 800 730-2215.)

This activity and others in this article were chosen because they represent, in several ways, many of the activities contained in these NSF-funded curricula; they do not stand above other choices. The reader also should be mindful of the fact that the extraction of an activity—one small slice of a unit—

from the surrounding instructional material conceals its rightful place in the full development of the mathematics in the curriculum. For example, the subsequent development surrounding the Gridville activity attends to "as a teacher" questions 4 and 5 below. As you reflect on your experience with this activity, consider these questions:

As a student:
1. Did you understand what you were being asked to do?
2. What mathematics did you do in order to answer the questions?

As a teacher:
1. What are the mathematical foci of this activity?
2. Is the mathematical content familiar?
3. Which Standards are embodied in this activity?
4. Without looking at any more material, do you feel that this activity could lead to further work for your less-capable students? How about your more-capable students?
5. Can you imagine students finding several different legitimate solutions to the problem? If yes, could such solutions lead in different directions? How would you move the entire class forward?

NEW FOCUS IN THE CLASSROOM

Let's consider the first "as a teacher" question and at least partially answer the second "as a student" question. According to the developers, the major thrust of this activity is for students to think and communicate about the mathematical modeling process. In particular, the major modeling emphases in this activity are the identification of key mathematical aspects of the context presented, the establishment of criteria (e.g., "What does 'best' mean?") against which potential solutions will be measured, and the simplification of a more general situation. The solution is not the main purpose of this opening activity. Finding, justifying, and analyzing solutions are objectives of the full unit.

Mathematics itself is a multidimensional discipline. It is a body of knowledge. It is a language. It contains algorithms and formulas. But if these are the only emphases in the classroom, then we have failed to concentrate on one of the most important aspects of the subject and we need to expand our focus. A primary characteristic of mathematics is that it provides a way of understanding and explaining phenomena: real, fanciful, or abstract; the tangible and typical "signatures" of mathematics are facts (theorems), algorithms, procedures, formulas, graphs, tables, or other representations. Stated succinctly, one of the central points of the *Standards,* and one of the most basic tenets of mathematics itself, is that mathematics is a sense-making activity. A salient feature of these NSF-funded curricula at every level is their emphasis on the sense-making activity of mathematics that requires reasoning and justification as part and parcel of understanding. Each level of these curricula is full of verbs such as *justify, demonstrate, explain, show, confirm, defend,* and so on. Moreover, the level of sophistication and degree of justification increases as students progress through the grades. In addition to considering the Gridville example, here is a short second example from the algebra strand in a middle school curriculum (Mathematics in Context, "Ups and Downs" unit, p. 53). In this unit, students are learning to represent and reason about growth patterns. These questions are suggested as individual informal assessments.

> Suppose you have two substances, A and B, that are changing in the following ways over the same length of time.
>
> Substance A: NEXT = CURRENT × 1/2
> Substance B: NEXT = CURRENT × 1/3
>
> 18. Which of the substances is changing more rapidly? Support your answer with a table or a graph.
> 19. Although one of the substances is changing more rapidly, both are decreasing in a similar way. How would you describe the way they are changing—faster and faster, linearly, or slower and slower? Explain.
>
> (Reprinted with permission from the Mathematics in Context® program, ©1999 Encyclopaedia Britannica, Inc.)

When most mathematicians talk about mathematical understanding, they are usually referring to a process that is much broader than providing the technical, logical details in the proof of a result. The process of mathematical understanding could entail these aspects: exploring examples and counterexamples; searching for patterns; drawing analogies; using intuition; using inductive reasoning; constructing mathematical models; posing problems; constructing and testing conjectures; justifying special cases; and finally, providing a deductive, general argument, possibly with additional assumptions or axioms, in order to obtain a mathematical result. Getting a feel for a mathematical situation and developing insight are extremely important parts of the process. Standard communication of a finished mathematical result in a theorem-proof format is intended to be an efficient method of communication; however, taken by itself, it can obscure many of the underlying aspects of the reasoning process. In fact, too often traditional school proofs (e.g., two-column proofs or proportional logic proofs) became almost an automated activity for students that led them away from a desire for real insight into a mathematical problem. A true search for proof usually requires exploration, creativity, cultivated intuition, and good judgment. Teachers may assume that students actually use many of these attributes to produce a technically accurate general proof. But this may not be the case

(Cuoco 1998). All students need explicit experience with the sense-making aspects of mathematical understanding, including what constitutes mathematical proof and the role of logic in mathematical certainty.

Which of these aspects of mathematical understanding do you experience when doing the high school example that accompanies figure 9.2?

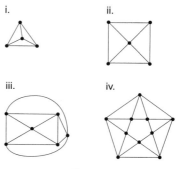

Fig. 9.2

1. Graphs have interesting properties that can be discovered by collecting data and looking for patterns.

a. Complete a table like the following one using the graphs above.

Graph	Sum of the Degrees of All Vertices	Number of Vertices of Odd Degree
i	30	
ii		2
iii		
iv		

b. Write down any patterns you see in the table.
c. Explain why the sum of the degrees of all vertices in *any* graph is an even number.
d. Explain why *every* graph has an even number of vertices with odd degree.

(From *Contemporary Mathematics in Context: A Unified Approach*, Part 1A, pp. 261–62. Chicago: Everyday Learning Corp., 1999.)

To facilitate students' comprehension of mathematics as a sense-making activity, teachers must understand and be able to convey the message that these aspects of mathematical understanding mark locations on an explore-conjecture-justify cycle. This fact is fortified when these aspects permeate a curriculum, as they do in these curricula.

Expanded View of Algorithms

Much of mathematics is concerned with algorithms. By an algorithm, we mean a procedure (usually a sequenced list of steps) guaranteed to produce a desired result. One of the operational characteristics in this definition is that not much thought is required to carry out the algorithm, other than possibly thinking about the order of steps in the procedure. That is, one of the primary objectives of an algorithm, or a formula, is to remove much of the need for thought. Although computational skill with algorithms and formulas is often important, a focus on such computational skill alone may promote a view of mathematics that directly opposes the view that mathematics is primarily a sense-making activity. Furthermore, computational skill is not necessarily an indicator of mathematical understanding. Yet, a lack of often-used computational skills can be an impediment to understanding when it sidetracks students, consuming their efforts, or when it leads students to erroneous results. Teachers must realize the importance of a balanced approach between pure computational skills and the sense-making activities outlined in the previous section.

The NSF-funded curricula include an expanded view of algorithms and formulas. Namely, algorithms and formulas often result from a thought process, as in the Gridville unit (where students develop and analyze at least two algorithms based on different optimization criteria), or they are a point of departure for further mathematical development or analysis, as occurs in the Growth example.

SETTING THE CONTEXT

Another feature found in these curricula is their use of problem-solving contexts as a means of introducing and developing mathematical content. These contexts may be real-world situations or purely mathematical settings, or they may be based on games, puzzles, or fantasy. Of course, one intention of a real-world context is to promote the value of mathematics. However, additional benefits result from the use of problem-solving contexts. *These contexts can create environments in which mathematical content is developed or mathematical understanding is enhanced.* A context can provide a local structure that adds coherence to a set of mathematical ideas and thought processes. Often these ideas are from different disciplines within mathematics. In this way, contexts are distinguished from other types of applications. For example, the Gridville problem to optimize the location of the fire station sets a context for the development of many mathematical ideas, including distance, circles, and ellipses in "fire truck" geometry (using a non-Euclidean metric in the plane), absolute value, functions and algebra involving the weighted sum of absolute functions, piecewise linear functions, and minimax solutions (choosing the minimum value in a set of several maximum values).

In a high school unit "High Dive" (*Interactive Mathematics Program,* Year 4 draft, p. 4, paraphrased below and accompanied by fig. 9.3), mathematics is developed and previous material extended while solving a unit problem:

> On a rotating Ferris wheel, a diver is being held by the ankles by an assistant on a platform that is attached to one of the seats on the Ferris wheel. A tub of water is

on a moving cart that runs along a track, parallel to the plane of the Ferris wheel, and passes under the end of the platform. The assistant must let go of the diver at exactly the right moment, so that the diver will land in the moving tub of water. Exactly when should the assistant release the diver?

(From *Interactive Mathematics Program Year 4*, published by Key Curriculum Press, 1150 6th Street, Emeryville, CA 94608, 1-800-995-MATH.)

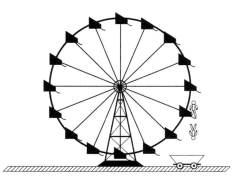

Fig. 9.3

This unit has an obvious focus on mathematical modeling and combines mathematical content from geometry, trigonometry, algebra, and calculus. Topics include periodicity and circular motion modeled by the sine and cosine functions, trigonometric identities, instantaneous velocity and the derivative, inverse functions, variation in functional parameters, quadratic expressions and the quadratic formula, polar coordinates, and expressions of velocity in terms of vertical and horizontal components. Topics from physics are also encountered.

A single contextual problem can last for an entire multiweek unit, as in High Dive, or several contexts may appear in a unit around a mathematical theme. Some mathematical development may occur beyond the limitations of a specific context, as in the Graph Theory example. Applications of mathematics after mathematical ideas have been developed are also found in these curricula.

Contexts can have additional pedagogical advantages. They can be wonderful motivators to capture students' interest. They can offer convenient opportunities to ask higher-order questions. Contexts also can serve as situations in which the mathematics is developed gradually, where students travel from the concrete to the abstract, or from a specific to a general situation. However, a rich context must not overwhelm or obscure the richness or the depth of the mathematical ideas. For example, students should be able to articulate clearly the mathematics that they encounter as they work through the Gridville activity.

MAKING CONNECTIONS

These curricula portray mathematics as a unified discipline by coherently integrating topics. Sometimes the integration of content is stratified, where topics from different mathematical disciplines are found in the same grade, but the material remains separate. At other times, integration involves amalgamating content. Real-world contexts often require this, as evidenced by High Dive and Gridville. In addition, all these curricula often tie mathematics to other disciplines. In fact, the elementary school curriculum Math Trailblazers is subtitled A Journey Using Science and Language Arts. The high school curriculum Integrated Mathematics: A Modeling Approach Using Technology provides choices in curricular paths based, in part, on variation in connections to disciplines outside mathematics.

Teachers need to be able to apply a variety of mathematical approaches to situations often involving several branches of mathematics as well as other disciplines. They must understand and be able to uncover connections in order to guide their students in their quest to extend mathematical understanding. As already seen, contexts are great ways to highlight or create ties between traditionally distinct mathematical disciplines.

CONTENT

Factual mathematical knowledge is so vast that it is impossible to include all worthy ideas in any grades K–12 curriculum in a meaningful way, and research suggests that we should not try to do so. Indeed, the Third International Mathematics and Science Study suggests that traditional school mathematics curricula contain too many topics (Schmidt, McKnight, and Raizen 1998). Moreover, there are several simultaneous goals for a curriculum designed for all students: to create mathematically literate citizens who recognize the cultural, intellectual, and pragmatic significance of mathematics; to offer mathematics relevant for the workplace; to create the capacity for independent learning; and to provide a foundation for further study in a variety of disciplines at the postsecondary level. In addition, technological advances affect the choice of curricular material. The curricular choices made by the developers of the NSF-funded instructional materials reflect all these conditions. Therefore, some traditional material has been discarded or coverage has been reduced, allowing for more in-depth exploration of a traditional topic or investigation of some new mathematical ideas.

It is not possible to give a complete content analysis here because mathematical topics and sequencing differ from curriculum to curriculum. See Robinson and Robinson (in preparation a, b, c) for details.

Implications for Teachers

Teachers at all levels need a solid foundation in statistics and probability. In elementary school statistics, topics include simple data collection, data displays, and data analysis. By high school, students are creating sophisticated sampling distributions and carrying out statistical analyses involving confidence intervals and margin of error. In elementary school, students determine simple experimental and theoretical probabilities. By high school, they progress to topics that involve some finite and continuous probability distributions as well as concepts such as conditional probability, independent events, and expected value. The data-collecting and data-analyzing strands play additional roles in integrating and unifying mathematical knowledge. Collecting data sometimes aids in statistical prediction. However, it is also pertinent to estimation, pattern recognition, the development of mathematical models through curve fitting, and the forming of conjectures based on the exploration of examples (a form of data collection).

Teachers at all levels need a solid foundation in geometry that includes some three-dimensional geometry, coordinate geometry, transformational geometry, symmetry, and a knowledge of both Euclidean and non-Euclidean geometry. Students begin to study two- and some three-dimensional Euclidean geometry early in elementary school. By the end of middle school, they have at least a basic understanding of the rigid geometric transformations, symmetry, congruence, similarity, area, and volume. By high school, students study Euclidean geometry in more depth as well as geometries such as elliptic, spherical, and hyperbolic.

Teachers at every level should have a solid background in calculus of a single variable. Students begin to encounter the area of irregular shapes in elementary school. This topic may be revisited with increasing sophistication; several high school curricula include some numerical integration. Rate of change is also a common theme and occurs at each level, in increasing complexity. (See the Growth example in this article for a middle school illustration.) In several curricula, students encounter the distinction between discrete and continuous quantities. In addition, some of the high school curricula contain a nonrigorous study of limits and the derivative.

Teachers at all levels need a good background in algebra and functions. In the elementary grades, the concept of variable and the use of symbols may be introduced. In the Everyday Mathematics curriculum, for example, fourth-grade students develop symbolic formulas for the area of some polygons. In the middle and high school curricula, matrices play an increasing role. Equation-solving techniques have been augmented to include some matrix and graphical methods, largely made possible by technology. Curve fitting is typical and may relate to linear functions, power functions, exponential functions, or quadratic functions, among others. Another frequent algebraic topic in the new curricula is the study of the graphical and contextual effects of changing the parameters A, B, C, and D in the expression $g(x) = Af(B(x - C)) + D$, where $f(x)$ is a known function. There is less emphasis on factoring and simplification of rational expressions. (See "Technology.")

Teachers at the middle and high school levels should have some familiarity with discrete mathematics. Topics that commonly appear in these curricula include combinatorics, recursive relations (e.g., the Growth example), linear programming, vertex-edge graph theory (e.g., Euler and Hamiltonian circuits and paths, graph coloring, and networks), scheduling, voting methods, fair allocation, and finite differences.

All the mathematical ideas and concepts contained in the union of these curricula, such as Arrow's theorem, Voronoi diagrams, Reuleaux polygons, the scan-conversion algorithm, and so on, are unlikely to be encountered in any teacher's formal education. Indeed, it should never be a goal of a teacher preparation program to cover all the content a prospective teacher is likely to encounter in a teaching situation. Rather, the goal should be to equip teachers with the ability to learn new mathematical content independently. In turn, teachers must be able to convey the same message to their students. Teachers also need to be able to facilitate work on extended problems, open-ended problems like Gridville, and problems with unknown solutions.

Pedagogical Perspectives

The NSF-funded curricula were based on research in teaching and learning supporting the proposition that a student will learn better and retain more when actively engaged in meaningful tasks (e.g., Lampert 1985; Lesh and Landau 1983; Schoenfeld 1987). Pedagogical approaches suggested in these curricula include work with students in groups, in pairs, and as individuals; class discussions; group and individual presentations; group and individual projects and homework; reflective or analytical writing; and direct instruction. Most of the curricula employ a variety of these instructional methods. Moreover, formative evaluation has indicated that lessons designed for use with one method may often be unsuccessful if approached using another strategy. Teachers thus need to know and be able to implement effectively a variety of pedagogical techniques in the classroom. Knowledge concerning the value and limitations of each strategy is crucial to successful use of this new generation of materials. Furthermore, pedagogy and the design of instructional materials are strongly linked. Recall "as a teacher" questions 4 and 5 following the Gridville activity.

Technology

If we consider the NSF-funded materials collectively, one can find the use of calculators, software drawing utilities, spreadsheets, data bases, specialized software programs, computer algebra systems, and the Internet to deliver the curriculum. The predominant technology is the calculator—four-function calculators in elementary school and graphing calculators in middle and high school. This is partly due to technology access and expense at the time these curricula were developed. However, several of these curricula use computers with varying frequency and ranging from optional use to extensively required.

All these curricula use technology in meaningful ways—ways that not only allow us to do things better but allow us to do better things. Technology is employed in these new materials to facilitate the observance of patterns and relationships, to create a virtual environment for exploration and conjecturing (as with a geometric drawing utility), to create simulations, to provide an effective means for using mathematical tools and operations (e.g., matrix multiplication or the computation of standard deviation), to implement some algorithms or procedures (e.g., determining a regression line), to access or organize data, to support a conjecture or general statement with experimental evidence, to check paper-and-pencil calculations, to facilitate the teaching of programming fundamentals, and to highlight the limitations of technology (e.g., a crucial error introduced by rounding, or the loss of information that occurs in digitized pictures). In addition, technology allows for the elimination or reduction in emphasis of some topics or skills, such as complicated long division done by paper and pencil. In one high school curriculum (Integrated Mathematics: A Modeling Approach Using Technology), a computer algebra system is used to simplify some rational functions, allowing release from tedious calculation and, more significantly, providing the opportunity to focus on the primary objective of the lesson: the concept of a linear asymptote. Technology also suggests new content such as computer graphics, dynamical systems, and fractals.

In summary, technology affects what students learn and how learning is accomplished. Teachers need to understand and be able to use technology in an ever-growing number of ways consistent with how people use it outside the classroom.

SUMMARY

The NSF-funded curricula at each school level have much in common in content, and the curricula overall have much in common in approach. We have attempted to elucidate some of their typical features and directions in the preceding pages. These features include a focus on mathematics as action undertaken by the student to make mathematical sense of situations, the use of contexts, the gradual development of content, the approach of moving from the concrete to the more abstract, students' development or analysis of algorithms as well as their implementation, the presentation of mathematics as a unified discipline, the systematic use of technology, and the inclusion of modern content.

Although these curricula have much in common, each supplies an individual answer to the question, "What should students know and be able to do at various levels in school mathematics?" It may be tempting to try to design a complete blueprint to be cloned for other textbook publishers by taking various intersections related to components of these *Standards*-based programs. But we believe it is more important to consider conclusions obtained from the union of these curricula. Taken together, these programs provide diversified ways to think "outside the box." Yes, one aspect of mathematics is that it is sequential, with results built on other results or assumptions and new knowledge that is dependent on previous knowledge. However, these curricula demonstrate that there are many reasonable and effective orderings of content topics that make curricular sense. For example, some content may appear earlier in a student's education than it did in more traditional programs. Other content, however, may be postponed until later, in a deliberate decision to position it closer to a subsequent topic. Differing selections of content can provide equally solid mathematical foundations for students. As is evidenced by the existence of these NSF-funded materials, it is possible to present coherent, integrated, multiyear curricula with considerable depth that are centered on problem situations and mathematical concepts yet focused on different mathematical questions.

Conclusion

It needs to be mentioned that there are other good instructional materials available; this specific collection of NSF-funded materials is one with which we are particularly familiar. And we have left much unsaid—for example, how these curriculum programs so artfully handle assessing students' work. However, the creation of instructional materials that truly support the *Standards* vision is only one step in the long process of reform in mathematics education. There are crucial implementation issues that go beyond the scope of this article.

As we strive to educate all students mathematically, we must recognize the need for both commonality and diversity. With this in mind, and building on existing curriculum models, there are several next steps and continuing challenges that could facilitate the development and implementation of curricula in the future. Can we, the wide mathematical community as well as parents and other stakeholders, agree on a small core of essential content knowledge that permits curricular choice and variety in a wealth of mathematical topics, covered in depth, in our classrooms? (Such an agreement, of course, would need periodic updating.) Can we agree on a small set of skills with which students should have reasonable facility and thus provide opportunity for immersion in the creative, thought-provoking world of mathematical exploration and reasoning? Can we create and maintain curriculum frameworks and high-stakes assessments that reflect these goals? Can we equip our current and future teaching force with this essential knowledge, the ability to learn and teach mathematical content with which they are not familiar, and the ability to facilitate students' mathematical exploration and reasoning? Can we further help a culture see beyond an apparently rigid façade to a world of mathematics teeming with possibilities, opportunities, and challenges for students now and in the future?

We predict that the twenty-first century will say that we must.

REFERENCES

Cuoco, Al. "What I Wish I Had Known about Mathematics When I Started Teaching: Suggestions for Teacher-Preparation Programs." *Mathematics Teacher* 91 (May 1998): 372–74.

Lampert, Magdelene. "How Do Teachers Manage to Teach?" *Harvard Educational Review* 55 (1985): 178–94.

Lesh, Richard, and Marcia Landau, eds. *Acquisition of Mathematics Concepts and Processes.* New York: Academic Press, 1983.

National Council of Teachers of Mathematics. *Assessment Standards for School Mathematics.* Reston, Va.: National Council of Teachers of Mathematics, 1995.

———. *Curriculum and Evaluation Standards for School Mathematics.* Reston, Va.: National Council of Teachers of Mathematics, 1989.

———. *Principles and Standards for School Mathematics: Discussion Draft.* Reston, Va.: National Council of Teachers of Mathematics, 1998.

———. *Professional Standards for Teaching Mathematics.* Reston, Va.: National Council of Teachers of Mathematics, 1991.

Robinson, Eric, and Margaret F. Robinson. "A Guide to Standards-Based Elementary Mathematics Instructional Materials." In preparation a.

———. "A Guide to Standards-Based Middle School Mathematics Instructional Materials." In preparation b.

———. "A Guide to Standards-Based Secondary Mathematics Instructional Materials." In preparation c.

Schmidt, William H., Curtis C. McKnight, and Senta A. Raizen. *A Splintered Vision: An Investigation of U.S. Science and Mathematics Education.* U.S. National Research Center for the Third International Mathematics and Science Study. East Lansing, Mich.: Michigan State University, 1996.

Schoenfeld, Alan H., ed. *Cognitive Science and Mathematics Education.* Hillsdale, N.J.: Lawrence Erlbaum Associates, 1987.

APPENDIX

NSF-Funded Instructional Development Materials

The following list includes the published name of each curriculum followed by a project name in parentheses that may be familiar to some readers. The grade levels of the curriculum materials are also included. Several of these curricula have been or are being published over several years. The date given in the reference relates to the latest publication date of portions of the materials together with an indication about what portion of the curriculum was published at the time this article was submitted.

Elementary School Instructional Development Projects

Everyday Mathematics, (EM or UCSMP elementary), K–6, University of Chicago School Mathematics Project, Everyday Learning Corporation, 1999.

Investigations in Data, Number and Space, K–5, TERC, Addison-Wesley Publishing Company, 1997.

Math Trailblazers: A Mathematical Journey Using Science and Language Arts, (TIMS), K–5, IMS Project University of Illinois at Chicago, Kendall/Hunt Publishing Company, 1998.

Middle School Instructional Materials Projects

Connected Mathematics, (CMP), 6–8, Glenda Lappan et al., Dale Seymour/Cuisenaire Publications, 1998.

Mathematics in Context, (MiC), 5–8, National Center for Research in Mathematical Sciences Education, Encyclopaedia Britannica, 1998.

MathScape, (STM), 6–8, Creative Publications, 1998.

Math Thematics, (STEM), 6–8, Rick Billstein, Jim Williamson, McDougal Littell Publishing Company, 1999.

Middle-School Mathematics through Applications Project, 6–8, Shelley Goldman et al., (MMAP), currently available through the Institute for Research on Learning (IRL).

Secondary School Instructional Materials Development Projects

Contemporary Mathematics in Context: A Unified Approach, (CPMP or Core-Plus), 9–12, Arthur F. Coxford et al., Everyday Learning Corporation (Courses 1–3), 1999.

Integrated Mathematics: A Modeling Approach Using Technology, 9–12, (SIMMS), Montana Council of Teachers of Mathematics and the Systemic Initiative for Montana Mathematics and Science, Simon & Schuster Custom Publishing Company, 1997. (Available from Pearson Custom Publishing Company)

Interactive Mathematics Program, (IMP), 9–12, Dan Fendel et al., Key Curriculum Press, (Years 1–3), 1998.

MATH Connections, a Secondary Mathematics Core Curriculum, (MATH Connections), 9–11, William P. Berlinghoff, Clifford Sloyer, Eric F. Wood, IT's ABOUT TIME, Inc., 1998.

Mathematics: Modeling Our World, (ARISE), 9–12, COMAP, Inc., South-Western Educational Publishing, (Courses 1–3), 1998.

10

Beyond Eighth Grade
Functional Mathematics for Life and Work

Susan L. Forman

Lynn Arthur Steen

For most of this century, schools have been organized to prepare some students for college and others for work. But the world of work is changing, much of it requiring increased use of data and computers, measurements and graphs. In the twenty-first century, most jobs and all careers will require some form of postsecondary education. Thus all students, regardless of career goals, will benefit from a curriculum that prepares them for both work and higher education. Instead of either/or, schools must now focus on both/and.

In this paper we propose a three-year high school program based on *functional mathematics* that prepares students for life and work. Typically, classroom applications are designed to serve mathematics—to offer context, to illustrate use, to motivate new concepts, or to integrate topics. In functional

---

Beginning in 1996 various industry groups published occupational skill standards to document the entry-level expectations of modern high-performance industries. At the same time the Institute on Education and the Economy at Teachers College, Columbia University, organized a series of meetings to stimulate discussion about integrating academic and industry skill standards. One such meeting, held in November 1997, focused on mathematics. "Beyond Eighth Grade" is one outcome of that meeting, a summary of issues and an interpretation of ideas that blend desires of employers with the expectations of academics.

The investigation that resulted in "Beyond Eighth Grade" was conducted under the auspices of the National Center for Research in Vocational Education (NCRVE), headquartered at the University of California at Berkeley. NCRVE was supported by the Office of Vocational and Adult Education, U.S. Department of Education. As such, this publication is in the public domain and may be reproduced as long as NCRVE and the report's authors are given full credit for its content. Points of view or opinions in this publication represent the views of the authors and do not necessarily represent official positions or policies of NCRVE or the U.S. Department of Education.

mathematics, the priorities are reversed. Instead of applications being introduced to help students learn a predefined body of knowledge, the mathematical topics are selected to prepare students to cope with common problems they will face in life and work. Examples of such real-life problems appear throughout this paper to illustrate in a concrete way the nature of functional mathematics.

Functional mathematics, if thoughtfully and rigorously developed, can provide a strong background both for students entering the work force and for those moving directly into postsecondary education. Indeed, for most students, functional mathematics will provide better preparation than current high school curricula. Functional mathematics stresses in-depth understanding of fundamental topics that are most likely to be used by large numbers of people. By employing concrete tools in settings that are both complex and realistic, functional mathematics pushes students to draw on the full breadth of mathematics. In short, focusing on useful mathematics increases total learning.

Given the complete record of a soccer league that ended in a three-way tie, devise a fair means of determining which team should be crowned champion.

NEED AND URGENCY

Despite mathematic's reputation as an ancient subject consisting of indisputable facts, mathematics education has recently become the source of passionate public debate. At stake is nothing less than the fundamental nature of school mathematics: its content (what should be taught), pedagogy (how it should be taught), and assessment (what should be expected). At times, these "math wars" have become so heated that Education Secretary Richard Riley has issued a public call for a truce (Riley 1998a).

At the risk of oversimplifying, this debate can be characterized as a clash between "traditionalists" who expect schools to provide the kind of well-focused mathematics curriculum that colleges have historically expected and "reformers" who espouse a broader curriculum that incorporates uses of technology, data analysis, and modern applications of mathematics. The reform approach is championed by the National Council of Teachers of Mathematics whose standards (NCTM 1989) advocate a robust eleven-year core curriculum for all students, with supplementary topics for those who are "college-intending." Critics argue, however, that the NCTM Standards are diffuse and ambiguous (Cheney 1997; Raimi and Braden 1998), that they are based on questionable assumptions about how students learn (Anderson, Reder, and Simon 1997), and that curricula based on these standards will not provide the kind of rigorous preparation students need to succeed in calculus and other college-level mathematics courses (Wu 1997).

Largely left out of this debate is a major constituency of mathematics education: employers. In today's information age, economic prosperity—for individuals as well as for the nation—depends on "working smarter, not just working harder." Yet a majority of America's businesses report deficiencies in the technical and problem-solving skills of their workers and a severe shortage of prospective employees with these requisite skills. The cost of finding skilled employees has become a serious impediment to growth in many sectors of the U.S. economy (Carnevale 1998).

Also left out are the voices of democracy and citizenship, which were so important in the development of public education in the United States. Now, two centuries later, quantitative literacy is every bit as important as verbal literacy for informed participation in civic affairs. Today's news is not only grounded in quantitative issues (e.g., budgets, profits, inflation, global warming, weather probabilities) but is also presented in mathematical language (e.g., graphs, percentages, charts).

Neither traditional college-preparatory mathematics curricula nor the newer standards-inspired curricula were designed specifically to meet either the technical and problem-solving needs of the modern work force or the modern demands of active citizenship. Although each includes much that is of value for citizenship and employment, neither provides the context, motivation, or balance of mathematical topics necessary for citizens or prospective employees in a data-drenched world.

The common curricular alternative—vocational or consumer mathematics—is significantly worse. Historically, vocational mathematics has provided only a narrow range of skills limited to middle school topics and devoid of conceptual understanding (National Center for Education Statistics 1996). Such programs leave students totally unprepared—not only for modern work and postsecondary education, but even for advanced secondary school mathematics. Although some innovative school-to-career programs are seeking to change this pattern of low expectations, the vast majority of secondary schools in the United States offer students no effective option for mathematics education that meets the expectations of today's high-performance workplace.

> *A student plans to take out a $10,000 loan at 7% interest with monthly payments of $120, but before she closes the deal interest rates rise to 7.5%. What will happen if she keeps her monthly payments at $120?*

EMPLOYMENT AND EDUCATION

Mathematics is the key to many of the most secure and financially rewarding careers in every sector of the economy (Business Coalition for Education Reform 1998). The impact of computers and information technology can be seen not just in engineering and science, but in such diverse areas as manufacturing and agriculture, health care and advertising. To be prepared for careers in virtually any industry, and especially for changing careers during a lifetime, secondary school students need to learn a substantial core of mathematics. However, this core is like neither the abstract pre-engineering mathematics of the academic curriculum nor the restricted topics of the discredited "vocational math." New approaches are needed to meet today's challenges.

A recent survey of 4,500 manufacturing firms revealed that nearly two out of three current employees lack the mathematics skills required for their work, and that half lack the ability to interpret job-related charts, diagrams, and flowcharts (National Association of Manufacturers 1997). Other reports cite a major shortage of qualified candidates for jobs in the information technology industries (Information Technology Association of America 1997), as well as for technicians and licensed journeymen in the skilled trades (Mathematical Sciences Education Board 1995). Even office work has changed so that technical skills are now at a premium (Carnevale and Rose 1998).

What current and prospective employees lack is not calculus or advanced algebra, but a plethora of more basic quantitative skills that could be taught in high school but are not (Murnane and Levy 1996; Packer 1997). They need statistics and three-dimensional geometry, systems thinking and estimation skills. Even more important, they need the disposition to think through problems that blend quantitative data with verbal, visual, and mechanical information; the capacity to interpret and present technical information; and the ability to deal with situations when something goes wrong (Forman and Steen 1998).

Business has discovered, and research confirms, that diplomas and degrees do not tell much about students' actual performance capabilities. For example, data from the National Assessment of Educational Progress (NAEP) (1997b) show that twelfth-grade students at the 10th percentile are essentially similar to fourth-grade students at the 80th percentile. Indeed, the level that NAEP considers "advanced," and which is achieved by only 8% of U.S. students, is considered just barely adequate in the context of college expectations (NAEP 1997a). Enrollment data for postsecondary mathematics courses confirm this discrepancy (Loftsgaarden, Rung, and Watkins 1997): three out of every four students enrolled in college mathematics courses are studying subjects typically taught in high school or even middle school (see fig. 10.1). Clearly, covering mathematics in school is no guarantee of mastering it for later use.

Nearly two-thirds of high school graduates enter postsecondary education primarily in order to obtain further skills and an advanced degree. Unfortunately, fewer than half of those who begin college attain any degree at all within five years. Furthermore, the majority of those who begin a traditional liberal arts program never finish. Although the economy clearly needs employees with advanced technical training (Judy and D'Amico 1997), these students—the majority—end up with just a list of courses and no degree or job certification (Barton 1997).

Ever since publication of *A Nation at Risk* (National Commission on Excellence in Education 1983), many advocates of educational reform have built their case on international competitiveness: to compete in a global economy that is increasingly technological, U.S. workers

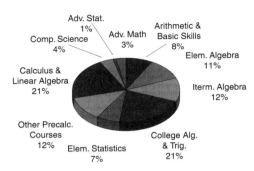

Fig. 10.1. 1995 Postsecondary Mathematics Enrollments

need better technical education (Commission on the Skills of the American Workforce 1990). Yet data from international comparisons such as the Third International Mathematics and Science Study (TIMSS) show that U.S. students are far from competitive (National Center for Education Statistics 1998). Thus, according to this argument, to remain internationally competitive we need to radically overhaul mathematics and science education (Riley 1998b).

In fact, the U.S. economy is thriving despite consistently weak performance by students on both national and international tests. This paradox has led some observers to suggest that the problem with weakness in school mathematics and science education is not so much that it hurts the overall economy, but that it increases economic inequities by providing the means to a good livelihood to only a few, primarily those from upper socioeconomic backgrounds (Barton 1997; Bracey 1997). From this perspective, the primary rationale for improving school mathematics is not competitiveness, but equity: in today's data-driven world, there is no justification for approaches to mathematics education that filter out those with greatest need and equip only the best-prepared for productive high-income careers.

A high school curriculum that helps all students master functional mathematics would effectively address issues of both equity and competitiveness. Since all students would study the same curriculum, all would have equal opportunity to master the mathematics required for the new world of work. Moreover, a three-year core of functional mathematics would give all students a strong platform on which to build either technical work experience or advanced education. Either route would lead to productive careers.

A large load of topsoil forms a conical pile. Because of its size, you cannot directly measure either its diameter or its height. Find a strategy for estimating its volume.

Theory and Practice

Historically, education in the United States has vacillated between the liberal and the pragmatic, between Robert Maynard Hutchins and John Dewey. Mathematics reflects a similar tension in the delicate balance of theory and practice, of the pure and the applied (Thurston 1990). Through most of this century, school mathematics has oscillated back and forth between these poles (Kilpatrick 1997). Indeed, nearly a century ago, the president of the American Mathematical Society lamented the "grievous" separation of pure from applied mathematics and urged schools to provide a more "practical" mathematics education: "With the momentum of such [education], college students would be ready to proceed rapidly and deeply in any direction in which their personal interests might lead them" (Moore 1903). Today's effort to make mathematics more functional for all students is just the latest chapter in this long saga.

In recent years this debate has been expressed in the form of standards, both academic and occupational. Coordinating these standards will involve not only issues of content and pedagogy, but also the balance of school-based vs. work-based learning (Bailey 1997). Historically, vocational curricula designed to prepare students for work have been burdened by second-class status in comparison with more rigorous academic curricula. Too often, vocational programs became dumping grounds for students who appeared slow or unmotivated—"other people's children." Most programs responded by limiting goals and lowering expectations, thereby offering stunted education to students who were already behind. In contrast, contemporary career-oriented curricula have been designed not primarily as training for low-skill jobs but as motivation for rigorous study, both academic and vocational (Bailey and Merritt 1997; Hoachlander 1997). By setting high standards, these programs offer significant responses to the twin challenges of equity and competitiveness.

Mathematics provides a microcosm of the duality between the academic and the vocational. Widely perceived as the epitome of theory and abstraction, mathematics is also valued as a powerful, practical tool (Odom 1998). In many occupations, quantitative literacy is as important as verbal literacy (Steen 1997); however, if mathematics education is to serve the world of work, a different type of experience than that found in typical mathematics courses is required (National Research Council 1998).

Between theory and application lies professional practice—the synthesis of thought and action employed by practitioners in all vocations. Many have argued that practice, properly understood, can be a legitimate and unifying goal of education. Practice is functional knowledge, the kind of know-how that allows people to get things done. According to educator Lee Shulman (1997), practice can provide a context in which theory becomes meaningful, memorable, and internalizable. Peter Denning (1997), a computer scientist, believes that practice—not knowledge or literacy—is what constitutes true expertise. Indeed, practice is what people tend to expect of schools, especially of mathematics education. It is at the heart of functional mathematics.

An infusion of practice into school mathematics can overcome what Shulman identifies as major deficiencies of theoretical learning: loss of learning ("I forgot it"), illusion of learning ("I thought I understood it"), and uselessness of learning ("I understand it but I can't use it"). Adults who are not professional users of mathematics will recognize these deficiencies from their own experiences. Little of what adults learned in school mathematics is remembered or used, so the accomplishment of "learning" mathematics is often an illusion. In fact, the mathematics many students are force-fed in traditional school environments creates a severe psychological impediment to the practice of mathematics in adult life (Cockroft 1982; Buxton 1991). Functional mathematics avoids many of these pitfalls by emphasizing that the goal of mathematics education is not just mathematical theory and word problems, but authentic mathematical practice.

> *Habitat for Humanity uses volunteer labor to build inexpensive homes, which it sells for the cost of materials. Using information on standard building supplies obtained from a local lumberyard, design a simple home whose building materials can be obtained for $15,000.*

HIGH SCHOOL MATHEMATICS

Traditionally, high school mathematics has served two different purposes —to prepare college-intending students for calculus (and other mathematics-based courses) and to equip other students with necessary skills, mostly arithmetic, so that they can function as employees, homemakers, and citizens. Although most traditionalists—and most parents and grandparents—still support these dual goals, reformers argue for a common curriculum for *all* students which emphasizes problem solving, communication, reasoning, and connections with other disciplines.

Proposed goals for school mathematics can be found in many sources. Some focus directly on K–12, others on the needs of postsecondary education or employers. The National Council of Teachers of Mathematics (NCTM 1989) provides a comprehensive set of standards for grade levels K–4, 5–8, and 9–12 that represents the "reform" perspective. In contrast, California recently adopted mathematics standards that represent a more traditional perspective (California Academic Standards Commission 1997). The American Mathematical Association of Two-Year Colleges articulated standards for college mathematics before calculus (AMATYC 1995) that include expectations for the mathematical foundation that students need to succeed in college. In addition, in the influential report *What Work Requires of Schools* (Secretary's Commission on Achieving Necessary Skills [SCANS] 1991), the U. S. Department of Labor outlined both foundation skills and broad employability competencies for mathematics and other subjects.

These standards differ greatly in both mathematical content and rhetorical style (see Appendix A), although most have overlapping goals. Indeed, to succeed in the real world of teachers and parents, schools and school boards, a mathematics curriculum must

1. meet society's expectations of what all high school graduates should know and be able to do;
2. reflect priorities common to state and national guidelines;
3. increase the number of students who successfully persist in advanced mathematics-based courses, including calculus;
4. enable students to see and use mathematics in everyday aspects of life and work;
5. help students understand and use correct mathematical language.

Functional mathematics must also meet these objectives. The first two establish priorities: to focus early and often on what everyone agrees must be learned, leaving to later (or to optional strands) those topics that only some students will find interesting or important. The third objective establishes a standard of quality: to increase the number of students who persist in further mathematics-based courses (including calculus, the traditional hallmark of mathematical success). The fourth objective conveys a commitment to utility—to ensure that students see mathematics as something real in their lives rather than as an alien subject encountered only in school. Finally, the fifth objective stresses command of the language of mathematics, a skill at least as important for success as a command of English.

By meeting these objectives, functional mathematics will satisfy the general public's expectations of school mathematics. In addition, these objectives also enhance functional mathematics' primary goal of preparing students for life and work. Consistent quality and high standards are essential in today's high-performance industries. Persistence in mathematics is not just of academic importance; it is also one of the best predictors of success in careers (Commission on the Skills of the American Workforce 1990). Moreover, the language of mathematics provides the power to analyze and express complex issues in all aspects of life and work. Fluency in this language is important not only for productive employees but also for careful consumers and critical citizens.

In functional mathematics, utility is center stage. Other objectives play important but supporting roles. Unfortunately, many mathematicians and mathematics teachers find utility at best a bleak justification (Howe 1998) for a subject that they chose for its beauty and elegance. For them, the power of mathematics—in Eugene Wigner's famous phrase, its "unreasonable effectiveness"—is not its primary virtue, but merely a consequence of its elegance and internal structure. Thus mathematicians are wont to stage their subject with theory and abstraction at the center, employing applications, technology, and practice as needed to help promote understanding.

To engage mathematicians and mathematics teachers, functional mathematics needs to be seen in terms of both utility and beauty. For many students, utility can be a path to beauty while for others, mathematics by itself provides sufficient internal motivation to sustain interest and accomplishment. For any mathematics curriculum to succeed with all students, it must build on the twin foundations of utility and elegance.

> *What measurements do you need to take in order to tile the floor of a room? How can you use these measurements to determine the number of regular tiles, border tiles, and corner tiles that are needed? What if you decide to lay the main tiles on a 45° angle?*

FUNCTIONAL MATHEMATICS

Functional mathematics comprises content, curriculum, context, and pedagogy. By content we simply mean the mathematics students should know and be able to do after finishing the first three years of high school mathematics (see Appendix B). Because mathematics is mathematics—whether traditional, reform, or functional—most of these elements are unsurprising. Although some topics are uncommon (e.g., index numbers, tolerances, three-dimensional geometry, normal curve, quality control charts, standards of proof, financial mathematics, spreadsheets), most are taught in any high-quality high school mathematics program. To achieve its goal of preparing students for both work and further education, functional mathematics respects the many parts of the traditional curriculum that are broadly useful, even while reshaping the boundaries to reflect its distinctive objectives.

The outline of functional mathematics in Appendix B reflects an inventory of mathematical topics selected for their importance in daily life and modern jobs as well as for their value in providing a strong foundation for further education. This outline is organized in predictable strands that cover what is normally subsumed under the umbrella of mathematics: numbers and data, measurement and space, growth and variation, chance and probability, reasoning and inference, variables and equations, modeling and decisions. Real problems cut across all this mathematics, just as these topics cut across the diverse contexts of authentic mathematical practice.

Clearly, many of the elements of functional mathematics are identical to the mathematics found in both traditional and reformed curricula. The core of school mathematics is more or less the same, even if viewed (or taught) from different perspectives. Percentages and ratios; linear and quadratic equations; areas, angles, and volumes; and exponential growth and trigonometric relations must be included in any strong high school mathematics program. The distinctions among traditional, reformed, and functional curricula lie not so much in core content as in contexts, emphases, and pedagogy.

Nonetheless, prospective employees for the new high-performance workplace need expertise in several aspects of mathematics not now emphasized sufficiently in school. On the one hand, students need greater experience recognizing and using some parts of middle school mathematics such as ratio, percentage, and measurement geometry that, although covered in current programs, are not used sufficiently to be learned well. On the other hand, as prospective employees, they need to understand and be able to use mathematical notions such as data analysis, statistical quality control, and indirect measurement that are hardly ever required in high school (Forman and Steen 1998).

In addition to shifting the balance of topics, functional mathematics provides much greater emphasis on "systems thinking"—on habits of mind that recognize complexities inherent in situations subject to multiple inputs and diverse constraints. Examples of complex systems abound—from managing a small business to scheduling public transportation, from planning a wedding to reforming social security. At all levels from local to national, citizens, policy makers, employees, and managers need to be able to formulate problems in terms of relevant factors and design strategies to determine the influence of those factors on system performance. Although such systems are often so complex that they obscure the underlying mathematics, the skills required to address realistic problems very often include many that are highly mathematical.

A curriculum built on functional mathematics gives students many opportunities to solve realistic problems and build mathematical understanding. Nevertheless, to make this learning valuable for work and further education—as well as to enhance understanding—such a curriculum must also help students become fluent in the language of mathematics. Individuals need to be able to read, understand, and interpret technical material with embedded charts and diagrams (e.g., property tax bills, stock market reports); they need to be able to speak clearly about mathematical ideas (e.g., as a salesman explaining the interest and payoff on an insurance policy); and they need extensive experience writing reports based on mathematical and technical analysis (e.g., a recommendation to a supervisor summarizing the conclusion of a study).

Functional mathematics channels the much-criticized "mile-wide, inch-deep" curricular river into a narrower but deeper stream of ideas and procedures that reinforce each other as students progress through school and college and on into careers. It provides a rich foundation of experience and examples on which students can build subsequent abstractions and generalizations. Indeed, to fulfill its goals, a functional curriculum must leave students well prepared not only for work but also for subsequent courses in more abstract mathematics.

> You are helping your brother-in-law build a garage on gently sloped land next to his house. After leveling the land, you begin staking out the foundation. To check that corners are square, you measure the diagonals and discover that they differ by 3 inches. Is that because the corners may not be perfectly level, or because they are not perfectly square? How can you determine what needs fixing to make sure that you start with a foundation that is both level and square?

FUNCTIONAL CURRICULA

The elements of functional mathematics can be embedded in many different curricula—the paths students follow through their education. Although some parts of mathematics impose a necessary order on the curriculum (e.g., arithmetic before algebra, linear equations before quadratic), large parts of mathematics can be approached from many different directions. Data analysis can be either a motivation for or an application of graphing and algebra, geometry can either precede or follow algebra, and each can reinforce the other. The order in which elements are listed bears no relationship to the order in which they may be taught through a three-year core curriculum.

The elements of functional mathematics arise from common contexts of life and work—measuring objects, managing money, scheduling time, making choices, and projecting trends. Although it is possible to organize a curriculum around such contexts, without a list of elements such as those in Appendix B to guide instruction, the mathematics itself may remain largely hidden. Alternatively, a functional curriculum can be organized around mathematical themes such as the sections of Appendix B. Indeed, the latter fits better the experience of most mathematics teachers and is more likely to be adaptable to most school settings.

Any mathematics curriculum designed on functional grounds—whether organized around external contexts or mathematical themes—will emphasize authentic applications from everyday life and work. In such a curriculum, students will gain considerable experience with mathematical tasks that are concrete yet sophisticated, conceptually simple yet cognitively complex (Forman and Steen 1995). A functional curriculum compels a better balance of statistics (numbers), geometry (space), and algebra (symbols)—the three major branches of the mathematical sciences. By highlighting the rich mathematics embedded in everyday tasks, this approach (in contrast to traditional "vocational math") can dispel both minimalist views about the mathematics required for work and elitist views of academic mathematics as an area with little to learn from work-based problems (Bailey and Merritt 1997; Forman and Steen 1998).

Because of the history of low standards in traditional vocational programs, many teachers and parents believe that a work-focused curriculum will necessarily lack the rigor of a pre-college academic track. Contrary to this belief, the "zero-defect" demands of the high performance workplace for exacting standards and precise tolerances actually impose a much higher standard of rigor than do academic programs that award students a B for work that is only 80–85% accurate. Moreover, the lengthy and subtle reasoning required to resolve many problems that arise in real contexts provides students with experience in critical thinking that is often lacking in academic courses that rush from topic to topic in order to cover a set curriculum.

> *Five friends meet for dinner in a restaurant. Some have drinks and others do not; some have dessert and others do not; some order inexpensive entrees, others choose fancier options. When the bill comes they need to decide whether to just add a tip and split it five ways, or whether some perhaps should pay more than others. What is the quickest way to decide how much each should pay?*

TEACHING FUNCTIONAL MATHEMATICS

Although the public thinks of standards primarily in terms of performance expectations for students, both the mathematics standards (National Council of Teachers of Mathematics 1989) and the science standards (National Research Council 1996) place equal emphasis on expectations for teaching, specifically that it be active, student-centered, and contextual:

- *Active instruction* encourages students to explore a variety of strategies; to make hands-on use of concrete materials; to identify missing information needed to solve problems; and to investigate available data.

- *Student-centered instruction* focuses on problems that students see as relevant and interesting; that help students learn to work with others; and that strengthen students' technical communication skills.

- *Contextual instruction* asks students to engage problems first in context, then with mathematical formality; suggests resources that might provide additional information; requires that students verify the reasonableness of solutions in the context of the original problem; and encourages students to see connections of mathematics to work and life.

These expectations for effective teaching are implicitly reinforced in recently published occupational skill standards (National Skills Standards Board 1998) that outline what entry level employees are expected to know and be able to do in a variety of trades. Although these standards frequently display performance expectations for basic mathematics as lists of topics, the examples they provide of what workers need to be able to do are always situated in specific contexts and most often require action outcomes (Forman and Steen, in press).

Most students learn mathematics by solving problems. In traditional mathematics courses, exercises come in two flavors: explicit mathematical tasks (solve, find, calculate, ...) and dreaded "word problems" in which the mathematics is hidden as if in a secret code. Indeed, many students, abetted by their teachers, learn to unlock the secret code by searching for key words (e.g., *less* for *minus*, *total* for *plus*) rather than by thinking about the meaning of the problem (which maybe a good thing because so many traditional word problems defy common sense).

In a curriculum focused on functional mathematics, tasks are more likely to resemble those found in everyday life or in the workplace than those found in school textbooks. Students need to think about each problem afresh, without the clues provided by a specific textbook chapter. Rather than just being asked to solve an equation or calculate an answer, students are asked to design, plan, evaluate, recommend, review, define, critique and explain—all things they will need to do in their future jobs (as well as in college courses). In the process, they will formulate conjectures, model processes, transform data, draw conclusions, check results, and evaluate findings. The challenges students face in a functional curriculum are often nonroutine and open-ended, with solutions taking from minutes to days, and requiring diverse forms of presentation (oral, written, video, or computer). As in real job situations, some work is done alone, and some in teams.

A round chimney 8" in diameter protrudes from a roof that has a pitch of 3:1. Draw a pattern for an aluminum skirt that can be cut out of sheet metal and bent into a cone to seal the chimney against rain.

MATHEMATICS IN CONTEXT

Students' achievements in school mathematics depend not only on the content of the curriculum and the instructional strategies employed by the teacher but also on the context in which the mathematics is embedded. Traditionally, mathematics has served as its own context: as climbers scale mountains because they are there, so students are expected to solve equations simply because it is in the nature of equations to be solved. From this perspective, mathematics is considered separate from and prior to its applications. Once the mathematics is learned, it supposedly can then be applied to various problems, either artificial or real.

Many of the new curricula developed in response to the NCTM standards or state frameworks give increased priority to applications and mathematical models. In some of these programs, applications are at the center, providing a context for the mathematical tools prescribed by the standards. In others, applications serve more to motivate topics specified in the standards. In virtually all cases, the applications found in current curricula are selected, invented, or simplified to serve the purpose of teaching particular mathematical skills or concepts. In contrast, the mathematical topics in a functional curriculum are determined by the importance of the contexts in which they arise.

For most students, interesting contexts make rigorous learning possible. Realistic problems harbor hidden mathematics that good teachers can illuminate with probing questions. Most authentic mathematical problems require multistep procedures and employ realistic data—which are often incomplete or inconsistent. Problems emerging from authentic contexts stimulate complex thinking, expand students' understanding, and reveal the interconnected logic that unites mathematics.

> *Devise criteria and procedures for fair addition of a congressional district to a state in a way that will minimize disruption of current districts while creating new districts that are relatively compact (nongerrymandered) and of nearly equal size.*

EMPLOYING COMPUTERS

It has been clear for many years that technology has changed priorities for mathematics. Much of traditional mathematics (from long division to integration by parts) was created not to enhance understanding but to provide a means of calculating results. This mathematics is now embedded in silicon, so training people to implement these methods with facility and accuracy is no longer as important as it once was. At the same time, technology has increased significantly the importance of certain parts of mathematics (e.g., statistics, number theory, discrete mathematics) that are widely used in information-based industries.

Many mathematics teachers have embraced technology, not so much because it has changed mathematics but because it is a powerful pedagogical tool. Mathematics is the science of patterns (Steen 1988; Devlin 1994), and patterns are most easily explored using computers and calculators. Technology enables students to study patterns as they never could before, and in so doing, it offers mathematics what laboratories offer science: a source of evidence, ideas, and conjectures.

The capabilities of computers and graphing calculators to create visual displays of data have also fundamentally changed what it means to understand mathematics. In earlier times, mathematicians struggled to create formal symbolic systems to represent with rigor and precision informal visual images and hand-drawn sketches. However, today's computer graphics are so sophisticated that a great deal of mathematics can be carried out entirely in a graphical mode. In many ways, the medium of computers has become the message of mathematical practice.

Finally, and perhaps most significantly, computers and calculators increase dramatically the number of users of mathematics—many of whom are not well educated in mathematics. Previously, only those who learned mathematics used it. Today many people use mathematical tools for routine work with spreadsheets, calculators, and financial systems—tools that are built on mathematics they have never studied. For example, technicians who diagnose and repair electronic equipment employ a full range of elements of functional mathematics—from number systems to logical inferences, from statistical tests to graphical interpretations. Broad competence in the practice of technology-related mathematics can boost graduates up many different career ladders.

This poses a unique challenge for mathematics education: to provide large numbers of citizens with the ability to use mathematics-based tools intelligently without requiring that they prepare for mathematics-based careers. Although mathematicians take for granted that learning without understanding is ephemeral, many others argue that where technology is concerned, it is more important for students to learn how to use hardware and software effectively than to understand all the underlying mathematics. But even those who only use the products of mathematics recognize the value of understanding the underlying principles at a time when things go wrong or unexpected results appear. In a functional curriculum where, for example, algebra emerges from work with spreadsheets, the traditional distinction between understanding and competence becomes less sharp.

—How many school teachers are there in New York City? How many electricians? How many morticians?

—How many words are there in all the books in the school library? How many megabytes of disk storage would be required to store the entire library on a computer?

AVOIDING PITFALLS

Those who develop materials and examples for a functional curriculum need to avoid some common pitfalls that plague all attempts at situating mathematics in authentic contexts. On the one hand, there is the temptation to give priority to the mathematics, either by selecting tasks to ensure coverage of mathematical topics without much regard to the tasks' intrinsic importance or by imposing unwarranted structure on a contextually rich problem in the interest of ensuring appropriate mathematical coverage. On the other hand, it is easy to overlook interesting mathematics hidden beneath the surface of many ordinary tasks or to choose problems that fail to help students prepare for advanced study in mathematics. Any curriculum that is to prepare students for subsequent mathematics-dependent courses must recognize the importance of intellectual growth and conceptual continuity in the sequencing of tasks in which mathematical activities are embedded.

Context-rich mathematics curricula often present tasks in the form of worksheets, outlining a series of short-answer steps that lead to a solution. While ostensibly intended to help students organize their thinking and assist teachers in following students' work, these intellectual scaffolds strip tasks of everything that makes them problematic. Indeed, worksheets reveal a didactic posture of traditional teaching (teacher tells, students mimic) that undermines learning and limits understanding. Students will learn and retain much more from the chaotic process of exploring, defending, and arguing their own approaches.

Finally, although a functional mathematics curriculum is motivated largely by examples that seem to lie outside the world of mathematics, it is nonetheless very important for students' future study that instructors bring mathematical closure at appropriate points. Students need to recognize and reflect on what they have learned; to be clear about definitions, concepts, vocabulary, methods, and potential generalizations; and to have sufficient opportunity to reflect on the accomplishments and limitations of mathematics as a tool in helping solve authentic problems.

> *A patient with an aggressive cancer faces two options for treatment: With Option A, he has a 40% chance of surviving for a year, but if he makes it that long then his chance of surviving a second year is two out of three. With Option B, he has a 50-50 chance of surviving each of the first and second years. Survival rates beyond the second year are similar for each option. Which choice should he make?*

MATHEMATICS IN LIFE AND WORK

The diverse contexts of daily life and work provide many realistic views of functional mathematics—of mathematical practice underlying routine events of daily life. These contexts offer episodic views, incomplete in scope and less systematic than a list of elements, but more suggestive of the way functional mathematics may be introduced in courses:

Reading Maps. Road maps of cities and states provide crucial information about routes and locations. For those who know how to "read" them, maps also convey scale and direction, helping drivers know which way to turn at intersections, permitting quick estimates of driving time, and revealing compass directions that relate to highway signs at road intersections. Map scales are just ratios—an essential part of school mathematics. Different scales not only convey different detail, but also require different translations to represent distance.

Reading maps is not just a matter of thinking of distances in different scales. In many cases, the geometry of maps represents other features such as temperature or soil content. Most common are weather maps with color-coded regions showing gradations in recorded or predicted temperatures. Similar maps sometimes display recorded or predicted precipitation, barometric pressure, vegetation features, or soil chemistry. Like topographic maps used by hikers, these maps represent some feature of the landscape that changes from place to place. The spacing between regions of similar temperature (or pressure, or elevation) conveys the steepness (or gradient) of change—what mathematicians call the "slope" of a line.

Scale-drawings and blueprints are also widely used to illustrate details of homes, apartments, and office buildings. These drawings represent sizes of rooms, locations of windows and doors, and—if the scale permits—locations of electrical outlets and plumbing fixtures. Architects' rulers with different units representing one foot of real space make it possible to read real distances off scale drawings, taking advantage of the geometrical properties of similar figures. New geographic information systems (GIS) encode spatially oriented data in a form suitable for computer spreadsheets, thereby enabling other factors (e.g., costs, environmental factors) to be logically linked to the geometric structure of a map.

Ensuring Quality. Statistical process control (SPC) and statistical quality control (SQC) are crucial components of high-performance manufacturing where "zero defect" is the goal. Instead of checking and repairing products after manufacture, firms like Boeing, General Motors, Kodak, Motorola, and Siemens now insist that at every step in the manufacturing process, materials, parts, and final products be manufactured within tight tolerances. Moreover, workers on assembly lines are responsible for ensuring this consistent level of quality.

The two tools that make this possible are based on statistics—the science of collecting and organizing data. The first, statistical process control, occurs during manufacture: assembly line workers chart key indicators of the process—perhaps the temperature of a mixture or the pitch of a grinding tool—on graph paper marked with curves representing limits determined by the required (or contracted) tolerances. If the process strays outside these limits, or approaches them too often, workers may decide to shut down the assembly line to make adjustments in the manufacturing process rather than risk manufacturing products that do not meet design specifications.

Statistical quality control is like statistical process control, but takes place when components (e.g., computer chips) are completed. By sampling finished products and charting their performance characteristics, workers can identify potential problems before products exceed permitted tolerances—and then take action to prevent the shipping or further manufacture of defective (i.e., out-of-tolerance) products.

Using Spreadsheets. Almost everyone who works with a computer uses a word processor for writing, whether for correspondence or business reports. Almost as popular are "number processors," commonly known as spreadsheets. Originally designed as a tool for accountants, spreadsheets are ubiquitous both in the office and at home—wherever anyone deals with budgets and expenses, taxes and investments. Spreadsheets are used to record business inventories and scientific data, to keep track of medical records and student grades, to organize crop records and airline schedules. Virtually any systematic information can be made more useful by being put in a properly organized spreadsheet.

To a mathematician, a spreadsheet is just algebra playing on a popular stage. The basic operations of a spreadsheet—adding cells together, calculating percentages, projecting growth rates, determining present values—are entered as formulas into the appropriate cells. More complex formulas (e.g., exponential, financial, trigonometric) are available from a pull-down menu. Once the computations are completed, the results can be displayed in graphs of various sorts (lines, bars, pies), often in vivid color.

Figuring out how to translate a task into a spreadsheet design is just like setting up a word problem in algebra: it involves identifying important variables and the relations among them. Preparing a spreadsheet requires equations which are suitably located in the cells. The spreadsheet does the arithmetic, and the designer does the algebra. Then, as in any mathematical exercise, the designer needs to check the results—typically by specifying independent computations to confirm key spots in the spreadsheet. (For example, adding all the entries in a grid can confirm the accuracy of the sum of the row totals, thus catching possible errors in the spreadsheet formulas.) Variables, equations, graphs, word problems—the ingredients of a good algebra course—are just the ticket for mastering spreadsheets.

Building Things. One in every four American workers builds things—automobiles or airplanes, bicycles or buildings, containers or chips. These products are three-dimensional, created by casting and cutting, by folding and fastening, by molding and machining. Designing things to be built (the work of engineers and architects) and building objects as designed (the work of carpenters and machinists) require impressive feats of indirect measurement, three-dimensional geometry, and visual imagination.

In a typical aluminum airplane part, for example, some measurements are specified by the designers while others must be calculated in order to program the cutting tool that will actually create the part. In three dimensions things are even more complicated. Planning how to drill holes at specified angles in a block of aluminum whose base is not square and whose sides are tilted in odd directions would tax the skills of most mathematics teachers. But machinists are expected to perform these calculations routinely to determine settings on a "sine plate," a device whose surface can tilt in two different dimensions in order to compensate for odd angles on the part that is to be drilled.

Both designers and builders now use computer-assisted design (CAD) and computer-assisted manufacturing (CAM) to ensure the exacting tolerances required for high-performance manufacturing. To use these tools effectively, workers need to have mastered basic skills of drawing geometric objects, measuring distances, and calculating angles, distances, areas, and volumes. The basic principles of geometry in three dimensions are the same as those in two dimensions, but the experience of working in three dimensions is startlingly more sophisticated. A good command of geometry and trigonometry is essential for anyone building things in today's manufacturing industries.

Thinking Systemically. Systems surround us—in commerce, science, technology, and society. In complex systems, many factors influence performance, thus making the task of solving problems inherently multidimensional. Indeed, the interaction of different factors is often difficult to predict, sometimes even counterintuitive. Complex systems defy simplistic single answers. Thus the first step in mathematical analysis is often to prepare an inventory of all possible factors that might need to be considered.

For example, the rise of efficient package delivery services and instantaneous computer communication have enabled many manufacturing companies to operate with minimum inventories, thus saving warehousing costs but risking a shutdown if any part of the network of suppliers fails. Understanding how a system of suppliers, communication, and transportation works requires analysis of capacity, redundancy, single-point failures, and time of delivery—all involving quantitative or logical analyses.

Other system problems arise within the everyday work of a typical small business. For example, the stockroom of a shoe store holds several thousand boxes labeled by manufacturer, style, color, and size, and arranged on floor-to-ceiling shelves. Deciding how to arrange these boxes can have a significant impact on the profit margin of the store. Obvious options are by manufacturer, by style, by size, by frequency of demand, or by date of arrival. Clerks need to be able to find and reshelve shoes quickly as they serve customers. But they also need to be able to make room easily for new styles when they arrive, to compare regularly the stockroom inventory with sales and receipt of new shoes, and to locate misshelved shoes. Mathematical thinking helps greatly in exploring the advantages and disadvantages of the many possible systems for arranging the stockroom.

Making Choices. Life is full of choices—to rent an apartment or purchase a home; to lease or buy a car; to pay off credit card debt or use the money instead to increase the down payment on a house. All such choices involve mathematical calculations to compare costs and evaluate risks. For example:

> The rent on your present apartment is $1200 per month and is likely to increase 5% each year. You have enough saved to put a 25% down payment on a $180 000 townhouse with 50% more space, but those funds are invested in an aggressive mutual fund that has averaged 22% return for the last several years, most of which has been in long-term capital gains (which now have a lower tax rate). Current rates for a 30-year mortgage with 20% down are about 6.75%, with 2 points charged up front; with a 10% down payment the rate increases to 7.00%. The interest on a mortgage is tax deductible on both state and federal returns; in your income bracket, that will provide a 36% tax savings. You expect to stay at your current job at least for 5–7 years, but then may want to leave the area. What should you do?

This sounds like a problem for a financial planner, and many people make a good living advising people about just such decisions. But anyone who has learned high school mathematics and who knows how to program a spreadsheet can easily work out the financial implications of this situation. Moreover, by doing it on a spreadsheet, it is very easy to examine "what if" scenarios: What if the interest rate goes up to 7% or 7.25%? What if the stock market goes down to its traditional 10–12% rate of return? What if a job change forces a move after three years?

In contrast to many problems of school mathematics which are routine for anyone who knows the right definitions (e.g., what is cos ($\pi/2$)?) but mystifying otherwise, this common financial dilemma is mathematically simple (it involves only arithmetic and percentages) but logically and conceptually complex. There are many variables, some of which need to be estimated; there are many relationships that interact with each other (e.g., interest rates and tax deductions); and the financial picture changes each year (actually, each month) as payments are made.

The complex sequence of reasoning involved in this analysis is typical of mathematics, which depends on carefully crafted chains of inferences to justify conclusions based on given premises. Students who can confidently reason their way through a lengthy proof or calculation should have no problem being their own financial advisors. And students who learn to deal with long chains of reasoning inherent in realistic dilemmas of life will be well prepared to use that same logic and careful reasoning if they pursue the study of mathematics in college.

In preparing fertilizer for a garden, a homeowner poured one quart of concentrated liquid fertilizer into a two-gallon can and filled the can with water. Then she discovered that the proper ratio of fertilizer to water should be 1:3. How much more liquid fertilizer should she add to the current mixture to obtain the desired concentration?

MAKING MATHEMATICS MEANINGFUL

Those who discuss mathematics education frequently describe mathematical knowledge in broad categories such as skills and understanding, concepts and facts, procedures and practices, or insights and knowledge. Whole volumes of educational research are devoted to distinguishing among these different aspects of mathematical knowledge. The standards movement has tended to subsume all these distinctions into two categories of knowledge and performance: what students should know and be able to do (Ravitch 1995; Tucker and Codding 1998).

The two broad cultures of mathematics education argue with each other largely because they differ in the interpretations they give to these different aspects of mathematical knowledge. Those who favor the traditional curriculum centered on algebra, functions, and Euclidean geometry argue that mastery of facts and basic skills are a prerequisite to understanding and performance. Reformers who favor a broader curriculum take a more constructivist view—that understanding and mastery are an outgrowth of active engagement with contextualized mathematics. Regardless of approach or emphasis, both traditional and reform curricula generally cover a similar set of topics designed to move students along the path from arithmetic to calculus.

Functional mathematics follows much the same path, with variations that reflect its grounding in authentic problems. However, by embedding mathematics in practice, functional mathematics can offer students both theory and know-how. Although in some technical areas, practical "of-the-moment" learning offers little that outlasts the next generation of gadgets, the logical structure that unites mathematics guarantees that all understanding, no matter how specific, has the potential to enhance mastery of other areas. What matters for long-term mastery of mathematics is not so much which particular skills are learned as that the process of learning be, in Shulman's (1997) words, "meaningful, memorable, and internalizable." Although topics in functional mathematics may be chosen for proximate utility, their study can provide insight and understandings sufficient for lifelong learning.

A curriculum based on functional mathematics requires appropriate content, authentic contexts, engaging tasks, and active instruction. By featuring mathematics in common contexts, a functional curriculum can motivate students to link meaning with mathematics. The best problem settings offer opportunities for exploration from multiple perspectives including graphical, numerical, symbolic, verbal, and computational. Technology—from graphing calculators and word processors to spreadsheets and symbolic algebra systems—can enhance understanding from each of these perspectives. Effective contexts provide opportunities for horizontal linkages among diverse areas of life and work as well as vertical integration from elementary ideas to advanced topics. Experience with rich contexts helps students recognize that asking questions is often as important as finding answers. Such contexts invite variations that can stimulate mathematical habits of mind and propel students to deep understanding.

Acknowledgements: While the ideas and proposals in this paper are the authors', we wish to thank the many colleagues and associates who have provided forceful arguments and stimulating critiques about the issues addressed in this paper. These include Thomas Bailey, Gene Bottoms, Maurice Burke, Gail Burrill, John Dossey, Teresa Drawbaugh, Rol Fessenden, Evelyn Ganzglass, Carver Gayton, Frank Girodano, Amy Gleason, Robert Glover, Amiee Guidera, Norton Grubb, Travis Hembree, Gary Hoachlander, Judith Leff, Jack Lochhead, Robert Kimball, Carole Lacampagne, Dane Linn, Charles Losh, Joyce Maddox, Kathy Mannes, Adrianne Massey, Pamela Matthews, James McKenney, Donna Merritt, Martin Nahemow, Robert Orrill, Arnold Packer, Jack Price, Larry Rosenstock, Rhonda Rumbaugh, Gerhard Salinger, Lisa Seidman, C. J. Shroll, Jimmy Solomon, Adria Steinberg, Elizabeth Teles, Margaret Vickers, Jack Wilkinson, and Joyce Winterton. We also thank Eric Larsen and Lisa Rothman for editorial assistance.

REFERENCES

American Mathematical Association of Two-Year Colleges. *Crossroads in Mathematics: Standards for Introductory College Mathematics Before Calculus.* Memphis, Tenn.: American Mathematical Association of Two-Year Colleges, 1995. Available: www.richland.cc.il.us/imacc/standards/.

Anderson, John R., Lynne M. Reder, and Herbert A. Simon. *Applications and Misapplications of Cognitive Psychology to Mathematics Education.* Pittsburgh, Pa.: Carnegie Mellon University, 1997. Available: act.psy.cmu.edu/personal/ja/misapplied.html.

Bailey, Thomas R. *Integrating Academic and Industry Skill Standards.* National Center for Research in Vocational Education. Berkeley, Calif.: University of California, 1997.

Bailey, Thomas R., and Donna Merritt. "School-to-Work for the College Bound." *Education Week* (29 October 1997): pp. 32, 37.

Barton, Paul E. *Toward Inequality: Disturbing Trends in Higher Education.* Princeton, N.J.: Educational Testing Service, 1997.

Bracey, Gerald W. *Setting the Record Straight: Responses to Misconceptions about Public Education in the United States.* Alexandria, Va.: Association for Supervision and Curriculum Development, 1997.

Business Coalition for Education Reform. *The Formula for Success: A Business Leader's Guide to Supporting Math and Science Achievement.* Washington, D.C.: U.S. Department of Education, 1998.

Buxton, Laurie. *Math Panic.* Portsmouth, N.H.: Heinemann, 1991.

California Academic Standards Commission. *Mathematics Standards.* Sacramento, Calif.: State Board of Education, 1997. Available: www.ca.gov/goldstandards/Drafts/Math/Math.html.

Carnevale, Anthony P. *Education and Training for America's Future.* Princeton, N.J.: Educational Testing Service, 1998.

Carnevale, Anthony P., and Stephen J. Rose. *Education for What? The New Office Economy.* Princeton, N.J.: Educational Testing Service, 1998.

Cheney, Lynne. "Creative Math or Just 'Fuzzy Math'? Once Again, Basic Skills Fall Prey to a Fad." *The New York Times* (11 August 1997): p. A13.

Cockroft, Sir Wilfred. *Mathematics Counts.* London: Her Majesty's Stationery Office, 1982.

Commission on the Skills of the American Workforce. *America's Choice: High Skills or Low Wages!* Rochester, N.Y.: National Center on Education and the Economy, 1990.

Denning, Peter J. "Quantitative Practices." In *Why Numbers Count: Quantitative Literacy for Tomorrow's America,* edited by Lynn Arthur Steen, pp. 106–17. New York: The College Board, 1997.

Devlin, Keith. *Mathematics: The Science of Patterns.* New York: W. H. Freeman, 1994.

Forman, Susan L., and Lynn Arthur Steen. *Beyond Eighth Grade: Report of a Workshop on Mathematical and Occupational Skill Standards.* New York: National Center for Research in Vocational Education, 1998.

———. "Making Authentic Mathematics Work for All Students." In *Education for Mathematics in the Workplace,* edited by Annie Bessot and James Ridgway. Dordrecht, Netherlands: Kluwer Academic, in press.

———. "Mathematics for Work and Life." In *Seventy-Five Years of Progress: Prospects for School Mathematics,* edited by Iris Carl, pp. 219–241. Reston, Va.: National Council of Teachers of Mathematics, 1995.

Hoachlander, Gary. "Organizing Mathematics Education Around Work." In *Why Numbers Count: Quantitative Literacy for Tomorrow's America,* edited by Lynn Arthur Steen, pp. 122–36. New York: College Board, 1997.

Howe, Roger. "Mathematics as Externality: Implications for Education." *Mathematics Education Dialogues* [National Council of Teachers of Mathematics] vol. no. 1 (March 1998): 8, 10–11.

Information Technology Association of America. *Help Wanted: The IT Workforce Gap at the Dawn of the New Century.* Arlington, Va.: Information Technology Association of America, 1997.

Judy, Richard W., and Carol D'Amico. *Workforce 2020: Work and Workers in the 21st Century.* Indianapolis, Ind.: Hudson Institute, 1997.

Kilpatrick, Jeremy. "Confronting Reform." *The American Mathematical Monthly* 104 no.10 (December 1997): 955–62.

Loftsgaarden, Don O., Donald C. Rung, and Ann E. Watkins. *Statistical Abstract of Undergraduate Programs in the Mathematical Sciences in the United States: Fall 1995 CBMS Survey.* Washington, D.C.: Mathematical Association of America, 1997.

Mathematical Sciences Education Board. *Mathematical Preparation of the Technical Work Force.* Washington, D.C.: National Research Council, 1995.

Moore, Eliakim Hastings. "On the Foundations of Mathematics." *Science* 217 (1903): 401–16.

Murnane, Richard, and Frank Levy. *Teaching the New Basic Skills: Principles for Educating Children to Thrive in a Changing Economy.* New York: Free Press, 1996.

National Assessment of Educational Progress. *The Mathematics Report Card.* Washington, D.C.: National Center for Education Statistics, 1997a.

———. *NAEP 1996 Trends in Academic Progress: Report in Brief.* Washington, D.C.: National Center for Education Statistics, 1997b.

National Association of Manufacturers. *The Skilled Workforce Shortage.* Washington, D.C.: National Association of Manufacturers, 1997. Available: www.nam.org/Workforce/survey.

National Center for Education Statistics. *Pursuing Excellence.* Washington, D.C.: U.S. Department of Education, 1998.

———. *Vocational Education in the United States: The Early 1990s.* Washington, D.C.: U. S. Department of Education, 1996.

National Commission on Excellence in Education. *A Nation At Risk: The Imperative for Educational Reform.* Washington, D.C.: U.S. Government Printing Office, 1983. Available: www.ed.gov/pubs/NatAtRisk/.

National Council of Teachers of Mathematics. *Curriculum and Evaluation Standards for School Mathematics.* Reston, Va.: National Council of Teachers of Mathematics, 1989. Available: www.enc.org/reform/journals/ENC2280/280dtoc1.htm.

National Research Council. *High School Mathematics at Work.* Washington, D.C.: National Academy Press, 1998.

———. *National Science Education Standards.* Washington, D.C.: National Academy Press, 1996.

National Skills Standards Board. *Occupational Skills Standards Projects.* Washington, D.C.: National Skills Standards Board, 1998. Available: www.nssb.org/ossp.html.

Odom, William E. *Report of the Senior Assessment Panel for the International Assessment of the U. S. Mathematical Sciences.* Washington, D.C.: National Science Foundation, 1998. Available: www.nsf.gov/cgi-bin/getpub?nsf9895.

Packer, Arnold. "Mathematical Competencies that Employers Expect." In *Why Numbers Count: Quantitative Literacy for Tomorrow's America,* edited by Lynn Arthur Steen, pp. 137–54. New York: College Board, 1997.

Raimi, Ralph A., and Lawrence S. Braden. *State Mathematics Standards.* Washington, D.C.: Thomas B. Fordham Foundation, 1998. Available: www.edexcellence.net/standards/math.html.

Ravitch, Diane. *National Standards in American Education: A Citizen's Guide.* Washington, D.C.: Brookings Institution, 1995.

Riley, Richard W. *Press Briefing, Feb. 24, 1988.* Washington, D.C.: U.S. Department of Education, 1998a.

———. "The State of Mathematics Education: Building a Strong Foundation for the 21st Century." Notices of the American Mathematical Society 45 (April 1998b): 487–90. Available: www.ams.org/notices/199804/riley.pdf.

Secretary's Commission on Achieving Necessary Skills (SCANS). *What Work Requires of Schools: A SCANS Report for America 2000.* Washington, D.C.: U.S. Department of Labor, 1991. Summary Available: www.stolaf.edu/other/extend/Resources/scans.html.

Shulman, Lee F. "Professing the Liberal Arts." In *Re-imagining Liberal Learning in America,* edited by Robert Orrill, pp. 151–73. New York: College Board, 1997.

Steen, Lynn Arthur. "The Science of Patterns." *Science* 240 (29 April 1988): 611–16.

———. *Why Numbers Count: Quantitative Literacy for Tomorrow's America.* New York: College Board, 1997.

Tucker, Marc S., and Judy B. Codding. *Standards for Our Schools: How to Set Them, Measure Them, and Reach Them.* San Francisco: Jossey-Bass, 1998.

Thurston, William P. "Mathematics Education." *Notices of the American Mathematical Society* 37, no. 7 (September 1990): 844–50.

Wu, Hung-Hsi. "The Mathematics Education Reform: Why You Should Be Concerned and What You Can Do." *The American Mathematical Monthly* 104, no. 10 (December 1997): 946–54.

APPENDIX A: EXCERPTS FROM DIFFERENT STANDARDS

The diverse ways that different organizations express expectations for mathematics illustrate a variety of approaches to setting standards. The excerpts that follow illustrate this variety in the particular case of algebra, the core of high school mathematics.

From the National Council of Teachers of Mathematics (1989, p.150):

In grades 9–12, the mathematics curriculum should include the continued study of algebraic concepts and methods so that all students can—

- represent situations that involve variable quantities with expressions, equations, inequalities, and matrices;
- use tables and graphs as tools to interpret expressions, equations, and inequalities;
- operate on expressions and matrices, and solve equations and inequalities;
- appreciate the power of mathematical abstraction and symbolism;

and so that, in addition, college-intending students can—
- use matrices to solve linear systems;
- demonstrate technical facility with algebraic transformations, including techniques based on the theory of equations.

From the California Academic Standards Commission (1997, pp. 76–80):

By the end of Grade 10 all students should be able to:

- Solve linear equations and inequalities with rational coefficients; use the slope-intercept equation of a line ($y = mx + b$) to model a linear situation and represent the problem in terms of a graph.
- Describe, graph, and solve problems using linear, quadratic, power, exponential, absolute value, polynomial, and rational functions; identify key characteristics of functions (domain, range, intercepts, asymptotes).
- Derive and use the quadratic formula to solve any quadratic equation with real coefficients; graph equations of the conic sections (parabola, ellipse, circle, hyperbola), identifying key features such as intercepts and axes.
- Describe, extend, and find the nth term of arithmetic, geometric, and other regular series.

And in Grades 11–12, mathematics students should learn about:

- piece-wise defined functions; logarithm function and as inverse of exponential; polar coordinates; parametric equations; recursive formulas, binomial theorem, mathematical induction; trigonometric functions, graphs, identities, key values, and applications; vector decomposition.

From the American Mathematical Association of Two-Year Colleges (1995, pp. 26–27):

The study of algebra ... must focus on modeling real phenomena via mathematical relationships. Students should explore the relationship between abstract variables and concrete applications and develop an intuitive sense of mathematical functions. Within this context, students should develop an understanding of the abstract versions of basic number properties ... and learn how to apply these properties. Students should develop reasonable facility in simplifying the most common and useful types of algebraic expressions, recognizing equivalent expressions and equations, and understanding and applying principles for solving simple equations.

Rote algebraic manipulations and step-by-step algorithms, which have received central attention in traditional algebra courses, are not the main focus.... Topics such as specialized factoring techniques and complicated operations with rational and radical expressions should be eliminated. The inclusion of such topics has been justified on the basis that they would be needed later in calculus. This argument lacks validity in view of the reforms taking place in calculus and the mathematics being used in the workplace.

From the Secretary's Commission on Achieving Necessary Skills (1991, pp. c1–c2):

Mathematics. Approaches practical problems by choosing appropriately from a variety of mathematical techniques; uses quantitative data to construct logical explanations for real-world situations; expresses mathematical ideas and concepts orally and in writing; and understands the role of chance in the occurrence and prediction of events.

Reasoning. Discovers a rule or principle underlying the relationship between two or more objects and applies it in solving a problem. For example, uses logic to draw conclusions from available information, extracts rules or principles from a set of objects or written text, applies rules and principles to a new situation, or determines which conclusions are correct when given a set of facts and a set of conclusions.

APPENDIX B: ELEMENTS OF FUNCTIONAL MATHEMATICS

These elements outline aspects of mathematics that are important for all students in their life and work. They emphasize concrete, realistic topics that arise in common situations in news, sports, finance, work, and leisure. These elements can be taught through many different curricula ranging from traditional to reform, from academic to vocational. Students completing any curriculum that includes these elements would be well-prepared to enter a wide variety of technical and academic programs, including a one-year precalculus course.

Numbers and Data

Mental Estimation. Anticipate total costs, distances, times; estimate unknown quantities (e.g., number of high school students in a state or city) using proportional reasoning; order of magnitude estimates; mental checking of calculator and computer results.

Numbers. Examples of whole numbers (integers), fractions (rational numbers), and irrational numbers (π, $\sqrt{2}$). Number line; mixed numbers; decimals, percentages, scientific notation. Prime numbers, factors; simple number theory; binary numbers and simple binary arithmetic; units and magnitudes; extreme numbers (e.g., national debt, astronomical distances); number sense; scientific notation.

Calculation. Accurate paper-and-pencil methods for simple arithmetic and percentage calculations; calculator use for complex calculations; spreadsheet methods for problems with a lot of data. Strategies for checking reasonableness and accuracy. Significant digits; interval arithmetic; errors; tolerances. Mixed methods (mental, pencil, calculator).

Coding. Number systems (decimal, binary, octal, hex); ASCII code; check digits. Patterns and criteria for credit card, Social Security, telephone, license plate numbers.

Index Numbers. Examples in the news: stock market averages; consumer price index; unemployment rate; SAT scores. Definitions and deficiencies; uses and abuses.

Information Systems. Collecting and organizing data; geographic information systems (GIS) and management information systems (MIS); visual representation of data.

Measurement and Space

Measurement. Direct and indirect means; estimation; use of appropriate instruments (rulers, tapes, micrometers, pacing, electronic gauges); plumb lines and square corners; calculated measurements; accuracy; tolerances; detecting and correcting misalignments.

Measurement Geometry. Measurement formulas for simple plane figures: triangles, circles, quadrilaterals. Calculation of area, angles, lengths by indirect means. Right triangle trigonometry; applications of Pythagorean theorem.

Dimensions. Linear, square, and cubic growth of length, area, volume. Coordinate notation; dimension as factor in multivariable phenomena.

Geometric Relations. Proof of Pythagorean theorem and of other basic theorems. Construction of line and angle bisectors, finding center of circular arc.

Spatial Geometry. Shapes in space; volumes of cylinders and spheres; calculation of angles in three-dimensions (e.g., meeting of roof trusses). Interpreting construction diagrams; nominal vs. true dimensions (e.g., of 2 × 4s); tolerances and perturbations in constructing three-dimensional objects.

Global Positioning. Map projections, latitude and longitude, global positioning systems (GPS); local, regional, and global coordinate systems.

Growth and Variation

Linear Change. Situations in which the rate of change is constant (e.g., uniform motion); contrast with examples where change is nonlinear (e.g., distance vs. time for falling body). Slope as rate of change; slope-intercept equation, with graphical significance of parameters. Difference between rate of change and value of the dependent variable.

Proportion. Situations modeled by similarity and ratio (height and shadows; construction cost vs. square footage; drug dose vs. body weight); examples where change is disproportional (e.g., height vs. weight). Calculating missing terms. Mental estimation using proportions.

Exponential Growth. Situations such as population growth, radioactivity, and compound interest, where the rate of change is proportional to size; doubling time and half-life as characteristics of exponential phenomena; symbolic representation (2^n, 10^n); ordinary and log-scaled graphs.

Normal Curve. Situations such as distribution of heights, of repeated measurements, and of manufactured goods in which phenomena distribute in a bell-shaped curve. Examples of situations in which they do not (e.g., income, grades, typographical errors, life spans). Parameters and percentages of normal distribution; z-scores, meaning of 1-, 2-, and 3-σ. Area as measure of probability.

Parabolic Patterns. Falling bodies; parabolas; quadratic equations; optimization problems.

Cyclic Functions. Situations such as time of sunrise, sound waves, and biological rhythms that exhibit cyclic behavior. Graphs of sin and cos; relations among graphs; $\sin^2 \theta + \cos^2 \theta = 1$.

Chance and Probability

Elementary Data Analysis. Measures of central tendency (average, median, mode) and of spread (range, standard deviation, mid-range); visual displays of data (pie charts, scatter plots, bar graphs, box and whisker charts). Distributions. Quality control charts. Recognizing and dealing with outliers. "Data = Pattern + Noise."

Probability. Chance and randomness; calculating odds in common situations (dice, coin tosses, lotteries); expected value. Binomial probability, random numbers, hot streaks, binomial approximation of normal distribution; computer simulations; estimating area by Monte Carlo methods. Two-way tables; bias paradoxes.

Risk Analysis. Common examples of risks (e.g., accidents, diseases, causes of death, lotteries). Ways of estimating risk. Confounding factors. Communicating and interpreting risk.

Reasoning and Inference

Statistical Inference. Rationale for random samples; double-blind experiments; surveys and polls; confidence intervals. Causality vs. correlation. Multiple factors; interaction effects; hidden factors. Judging validity of statistical claims in media reports. Making decisions based on data (e.g., research methods, medical procedures).

Scientific Inference. Gathering data; detecting patterns, making conjectures; testing conjectures; drawing inferences.

Mathematical Inference. Logical reasoning and deduction; assumptions and conclusions; axiomatic systems; theorems and proofs; proof by direct deduction, by indirect argument, and by "mathematical induction."

Verification. Levels of convincing argument; persuasion and counterexamples; logical deduction; legal reasoning ("beyond reasonable doubt" vs. "preponderance of evidence"; court decisions interpreting various logical options); informal inference (suspicion, experience, likelihood); classical proofs (e.g., isosceles triangle, infinitude of primes).

Variables and Equations

Algebra. Variables, constants, symbols, parameters; equations vs. expressions. Direct and indirect variation; inverse relations; patterns of change; rates of change. Graphical representations; translation between words and graphs. Symbols and functions.

Equations. Linear and quadratic; absolute value; 2×2 systems of linear equations; inequalities; related graphs.

Graphs. Interpretation of graphs; sketching graphs based on relations of variables; connection between graphs and function parameters.

Algorithms. Alternative arithmetic algorithms; flow charts; loops; constructing algorithms; maximum time vs. average time comparisons.

Modeling and Decisions

Financial Mathematics. Percentages, markups, discounts; simple and compound interest; taxes; investment instruments (stocks, mortgages, bonds); loans, annuities, insurance, personal finance.

Planning. Allocating resources; management information systems; preparing budgets; determining fair division; negotiating differences; scheduling processes, decision trees; PERT charts; systems thinking.

Mathematical Modeling. Abstracting mathematical structures from real-world situations; reasoning within mathematical models; reinterpreting results in terms of original situations; testing interpretations for suitability and accuracy; revision of mathematical structure; repetition of modeling cycle.

Scientific Modeling. Role of mathematics in modeling aspects of science such as acceleration, astronomical geometry, electrical current, genetic coding, harmonic motion, heredity, stoichiometry.

Technological Tools. Familiarity with standard calculator and computer tools: scientific and graphing calculators (including solving equations via graphs); spreadsheets (including presentation of data via charts); statistical packages (including graphical displays of data).

11

Statistics for a New Century

Richard L. Scheaffer

WE ARE surrounded by numbers, engulfed in them, perhaps even drowning under them. We live by the numbers, and the number crunch seems to be expanding rather than shrinking. Many of the numbers that regulate our lives, from movie ratings to measures of food quality, from the unemployment rate to the consumer price index, are gathered by others using complicated processes that are, seemingly, beyond our capacity to understand. We see from news reports, however, that there are problems with estimating unemployment rates, establishing a consumer price index, deciding which foods are healthy and which are not, or even counting the residents of the country. The numbers that surround these issues arise from processes that are understandable, at least in a general sense, by someone with a little knowledge of statistics. These same processes are used in business and industry to measure productivity, improve quality, and manage systems, so that a knowledge of statistics is essential for both good citizenship and productive employment. Colleges and universities have long recognized this fact; now almost all undergraduate majors require some basic knowledge of statistics. In recent years, this interest in, and enthusiasm, for statistics has moved into the grades K–12 mathematics curriculum, as can be seen in most state guidelines and the National Council of Teachers of Mathematics *Principles and Standards for School Mathematics* (NCTM 2000).

Conventionally, statistics has been taught as a series of techniques rather than a process of thinking about the world. Teachers and students tend to emphasize particulars rather than principles, narrow mechanics rather than broad methodologies, and specific formulas rather than general formulations. Techniques are useful, and perhaps that is where instruction in a discipline must begin, but now the instruction in and practice of statistics must move beyond the magical use of textbook or technological procedures to clear understanding of analyses and communication of results—beyond rote to reflection. At the introductory college level and, indeed, at the grades K–12 level, the guidelines set out by the ASA-MAA Focus Group (Cobb 1992) in the early 1990s provide a means to effect change in statistics education of the twenty-first century. These guidelines are built around the three-point foundation shown below.

- Emphasize statistical thinking
- Use more data and concepts, less theory and fewer recipes
- Foster active learning

Modern statistics education must have data analysis as its heart. Good analyses, however, involve careful thinking, and the whole statistical process is best learned in an active environment. (These are precisely the guidelines that allowed the AP Statistics program [College Board 1998] to get off to a successful start.) The next four sections of this article concentrate on showing how these guidelines can help broaden and deepen the teaching and learning in four important areas of statistics: (1) exploratory data analysis, (2) the fundamental concept of association, (3) inferential reasoning, and (4) principles of planning studies. The last section presents an overview of important statistical concepts that can provide a framework for teaching statistics in grades K–12 in the twenty-first century. Throughout the article, it is hoped that the usefulness of statistics as a way to encourage and illustrate many topics in mathematics will show through.

A CONNECTED AND DEEPER VIEW OF EXPLORATORY DATA ANALYSIS (EDA)

Exploratory data analysis has caught the imagination of teachers and students all the way down to the elementary grades, and no doubt is what most teachers have in mind when they say, "I teach a unit on statistics." It is fine to be able to construct a stem-and-leaf plot or a box plot, but learning must move beyond the level of construction to the levels of understanding and creative use. True developers and disciples of EDA speak of the four Rs: Revelation, Residual, Re-expression, and Resistance (Hoaglin and Moore 1992; Hoaglin, Mosteller, and Tukey 1983). Data usually reveal something if explored properly. The revelation may come about by looking at the departures (residuals) from an underlying pattern. Often, the scale of measurement must be changed (transformed) so that data can be re-expressed in a more useful way, or the rough spots in the data must be smoothed. The most useful summaries of the data are often those that resist a few anomalies in the measurements.

These points will be illustrated with data on the 50 states of the United States. The variables are the area of the state (thousands of square miles) and population density (people per square mile) according to 1990 census data. Figure 11.1 is a box plot of the population densities.

That the distribution of densities is highly skewed toward the larger values might be a revelation for those who have not thought about the variable; the fact that, by standard rules of operation, there are five outliers is quite surprising. Are all these small states really unusual? The question arises because a highly skewed distribution is expected to produce values in the tail that are far from the center; that's what makes it skewed. The usual definition of outlier is not appropriate here. Suppose, however, the data are re-expressed in units that are more natural for data sets that involve a few very large values and lots of small ones by taking the natural logarithm of each value. Figure 11.2 shows the box plot of the log-densities. Now, the distribution is nearly symmetric, except for one lone outlier, Alaska. Which of the 50 states could be more easily defended as being unusual, the five small northeastern ones, with high densities, or the huge state of Alaska, with a low density? The standard rule for identifying outliers works well when the data distribution is a somewhat symmetric. One of the first jobs of the data analyst, then, is to find a scale of measurement that allows the important features of the data to be easily seen and described.

Fig. 11.1. Population densities of the states, 1990

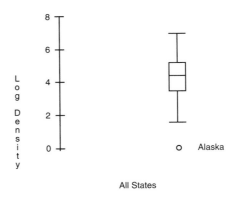

Fig. 11.2. Logarithms of population densities of the states

The importance of re-expression carries over to the issue of making comparisons. Figure 11.3 represents the densities for the states by region (1=Northeast, 2=Midwest, 3=South, and 4=West), whereas figure 11.4 represents the log-densities by region. Once the variation is stabilized a bit, the pattern of high densities in the East and low densities in the West is clear, with the Midwest and South being somewhat comparable to each other, and the real outliers show up (on the high side, Maryland in the South, California and Hawaii in the West; on the low side, Alaska in the West).

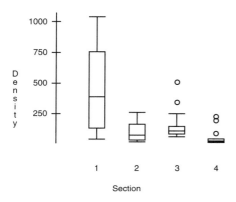

Fig. 11.3. State population densities by section of the country

If numerical measures of the differences among the regions are needed, it would be quite appropriate to compare the medians of the log-densities because the variations in these sets of measurements are similar and the medians resist the influence of the outliers that are still present.

Standard statistical procedures work best when the distribution of data is somewhat mound-shaped and symmetric—approximately normal.

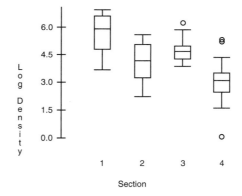

Fig. 11.4. Logarithms of state population densities by section for the country

But how do we tell when a distribution is approximately normal? Although there are complex statistical tests for normality, at the introductory level the best way to check normality is to produce a good picture of the data distribution. A histogram is the common starting place for investigating the shape of a distribution, but histograms tend to be too rough and are quite sensitive to the choice of bin width. A better picture can be obtained with a density estimate, which smoothes out a histogram over a variety of possible bin widths. Figures 11.5 and 11.6 show the estimated density curves (along with the histograms) for the data on population density of the states and the logarithm of density. In figure 11.5 the density curve tells more about the shape of the data than does the histogram. The data in the first bar tends to pile up around 0 with few observations around 200, for example, and there are a couple of data points around 1000. In figure 11.6 the density curve suggests a strong dose of normality.

The natural logarithm has served well as a transformation in the examples used above, but one might ask what other transformations prove to be useful.

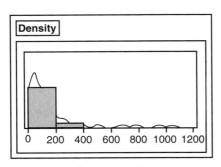
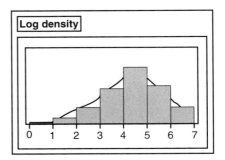

Fig. 11.5. Histogram and density curve of state population densities

Fig. 11.6. Histogram and density curve of logarithms of state population densities

In an introductory look at statistical concepts through basic EDA, logs, square roots, squares, and reciprocals can get the data analyst a long way. If students understood these basic four, they would be well on their way to a useful study of EDA. There is a handy rule for transformations that stabilize the standard deviations within groups so that comparisons of location can be made more appropriately. With s denoting standard deviation and $\bar{y}$ the mean, the rule says the following:

If s is proportional to $\bar{y}^{1-p}$ then make the transformation $\bar{y}^p$.

For the data on population densities by section of the country, the relationship between s and $\bar{y}$ yields $p = .2$, approximately. Transforming each of the density measurements to the .2 power results in parallel box plots very similar to those pictured in figure 11.4; the transformation does a good job of stabilizing the variation from region to region.

ASSOCIATION: A DEEPER LOOK AT RELATIONSHIPS

Fortunately, it turns out that re-expressions (transformations) that clarify shapes and allow meaningful comparisons are often the same ones that will simplify bivariate relationships. Suppose an investigator wants to see how well population density can be predicted from the areas of the states. This calls for a scatterplot and possibly a regression line relating area to density. Such a plot is shown in figure 11.7, but the relationship would require something much more complicated than a straight line as a model. However, if log-density is used as the response, the relationship "straightens out" some, as shown in figure 11.8. There is an underlying negative trend that could be linear except for the undue influence of the three big states, Alaska, Texas, and California, in decreasing order of area. If these influential points are removed, area is a decent predictor of log-density.

Finding possible associations between variables and measuring their strengths is one of the key ideas of statistics that flows throughout almost all EDA and statistical inference. Broadly speaking, data comes in two types: categorical and measurement. The sections of the country in which the states lie is a categorical variable, even though we used the numbers 1, 2, 3, and 4 to identify them in the earlier analyses, and the densities of people living in those states is a measurement variable. In an earlier section, we flirted with association when looking at the log-densities by section (measurement by categorical association). Figure 11.8 shows the association between log-density and area of the states (measurement by measurement association). To take these ideas a little deeper, consider scores and grades of students on the first two exams in an introductory statistics course (see table 11.1). Each score is a percentage between 0 and 100 but, to simplify things, scores are translated into grades so that each score in the 90s is an A, each score in the 80s is a B, and each score below 80 is a C (giving some students a break).

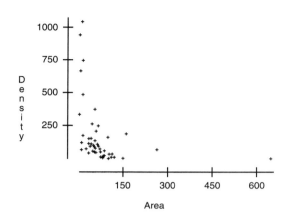

Fig. 11.7. Scatterplot of state population densities versus state areas

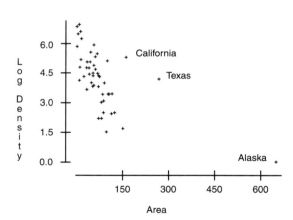

Fig. 11.8. Scatterplot of logarithms of state population densities versus state areas

Here is another table that shows the association between the grades on these two exams. Table 11.2 shows, for example, that 10 students received an A on both exams.

TABLE 11.1
Exam Score and Grade Data
(2 = C, 3 = B, 4 = A)

Exam 1	Exam 2	Grade 1	Grade 2
80	88	3	3
52	83	2	3
87	87	3	3
95	92	4	4
67	75	2	2
71	78	2	2
97	97	4	4
96	85	4	3
88	93	3	4
100	93	4	4
88	86	3	3
86	85	3	3
81	81	3	3
61	73	2	2
97	92	4	4
96	99	4	4
78	90	2	4
93	88	4	3
92	92	4	4
91	93	4	3
96	92	4	4
69	73	2	2
76	87	2	3
91	91	4	4
98	97	4	4
83	89	3	3
96	83	4	3
95	97	4	4
80	86	3	3

TABLE 11.2
Letter Grades for Exam 1 and Exam 2

	Exam 1		
Exam 2	C	B	A
A	0	4	10
B	0	7	1
C	4	2	1

A little inspection of the table shows that the large counts are on the main diagonal; those who get Cs on the first tend to get Cs on the second, and similarly with Bs and As. Thus, there is a positive association between the two sets of grades, where "positive" has the same interpretation as positive trend on a scatterplot. But how might we measure the strength of this association? Some will say, "Conduct a chi-square test." That can be done, and it is highly significant, but that tells us only that there is evidence of an association; it tells us nothing about the strength of the association or about its direction.

Many measures of strength are available, and some of the simpler ones should be introduced into elementary statistics, even at the middle or high school level. Most serious statistics packages compute them anyway. One that is easy to construct by hand is based on concordant and discordant pairs. Two students have concordant grades if one scored higher than the other on both exams—that is, an (A,A) pair is concordant with a (B,C) pair. Two students have discordant grades if Student 1 scored higher than Student 2 on one of the exams but the reverse was true on the other; that is, a (B,A) pair is discordant with an (A,C) pair. All ties on either exam are discarded. A measure of strength of the association can be formed by simply calculating the difference between the proportions of concordant and discordant pairs. For the table of exam grades, the 7 (B,B) pairs are concordant with each of the 4 (C,C) pairs, the 4 (B,A) pairs are concordant with the 4 (C,C) pairs and the 0 (C,B) pairs, and so on. Altogether, the number of concordant pairs, C, is given by C = 7(4) + 4(4+0) + 1(4+2) + 10(0+7+4+2) = 180. The number of discordant pairs, D, is given by D = 7(1) + 4(1+1) + 0 + 0 = 15. A measure of the strength of this association is then (C−D)/(C+D) = 165/195 = .85. There is a strong, positive association between the grade on the first exam and the grade on the second exam.

If the grades are now replaced by the actual scores, the appropriate plot to show a possible relationship is a scatterplot, as shown (with the regression line) in figure 11.9. The linear shape of the cloud of points has a central trend that is captured well by the regression line, and the correlation coefficient—.76 in this case—is a good measure of the strength of this association. The point on the extreme left produces a very large residual, and the strength of the relationship would be improved if this pair of scores were removed. (What

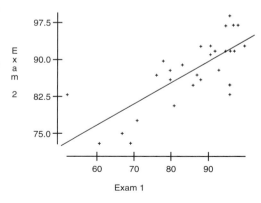

Fig. 11.9. Scatterplot of Exam 2 scores versus Exam 1 scores with regression line

is unusual about this pair of scores?) Again, a strong, positive association between scores on the first two exams is seen. Notice that the summaries have the same features whether the data is categorical (grades) or measurement (scores); a display (table or graph) that shows a linear trend is followed by a numerical measure of the strength of that trend.

Earlier, the roughness was smoothed out of a histogram to get a better picture of the distribution of data. Similarly, the roughness can be smoothed out of a scatterplot to show finer patterns in the data than can be seen from casual observation of the cloud of points. Figure 11.10 shows the same scatterplot and regression line as figure 11.9, but a "smoother" has been placed through the data. (The one used here is a *lowess*, or *locally weighted scatterplot smoother*.) On comparing the regression line to the smooth line, it is apparent that the linear model fits well on scores above 75, but a different line should be fit to the scores below 75 on Exam 1.

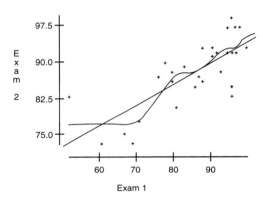

Fig. 11.10. Scatterplot of Exam 2 scores versus Exam 1 scores with regression line and smoother

There are two other types of associations that could be of interest with similar data on the scores and grades on two exams. If we had the scores (measurements) on Exam 2 but only the grades (categories) on Exam 1, we could plot the parallel box plots, as shown in figure 11.11, and then make comparisons of the medians or means, the latter lending itself to an introduction to the analysis of variance. Again, we see a strong, positive trend among the mean scores on Exam 2 as the grade on Exam 1 increases.

If, on the other hand, we had the scores (measurements) for Exam 1 and want to use them to predict a grade (category) for Exam 2, we are in a situation for which ordinary regression analysis will not work. The most common approach to solving this problem is through the use of logistic regression, which is now a widely used technique available in most standard com-

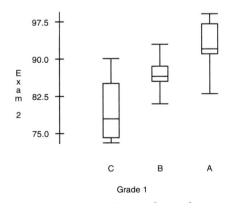

Fig. 11.11. Scores on Exam 2 by grades on Exam 1

puter packages for statistical analysis. In a logistic model the response of interest is the odds in favor of getting an A on Exam 2 and the explanatory variable the actual score on Exam 1. If the estimated probability of getting an A on Exam 2 is denoted by p, the logistic regression fits the model

$$\ln\left(\frac{p}{1-p}\right) = a + bx$$

where x denotes the score on Exam 1. So, the log odds are assumed to be linearly related to the explanatory variable, x. For the data on exam scores the solution is

$$\ln\left(\frac{p}{1-p}\right) = -17.26 + 0.19x$$

This implies that the odds of getting an A are related to first exam score by the model

$$\left(\frac{p}{1-p}\right) = \exp(-17.26 + 0.19x) = [\exp(-17.26)](1.21)^x$$

and we have the interesting result that the odds in favor of getting an A increase exponentially with the score on the first exam, going up 21 percent for every one-point increase in that score. A score of 80 on the first exam results in $p = .12$ as the estimated probability of an A on the second exam while a score of 95 on the first exam results in $p = .70$ as the estimated probability of an A on Exam 2. Logistic models can be fit to data even when the explanatory variable is categorical, but space does not permit that discussion here.

In summary, association between a response variable and an explanatory variable commonly falls into one of four classifications, as shown in table 11.3. An introduction to statistical thinking should encourage students to consider all four possibilities, and it should provide techniques for handling all four. After all, another maxim of modern data analysis says "an approximate solution to the correct problem is far better than an exact solution to the wrong problem." For far too long statistics education has been forcing many problems into an incorrect mold just so a standard solution could be used. It is high time to think more broadly.

TABLE 11.3
Relationships Between a Response Variable and an Explanatory Variable

Response Variable	Explanatory Variable	
	Categorical	Measurement
Categorical	Two-way tables Chi-square Measure of association	Logistic regression models based on log odds
Measurement	Parallel box plots Comparison of medians or means (ANOVA)	Scatterplots Regression Association (Correlation)

INFERENCE THROUGH SIMULATION AND RESAMPLING: THE BOOTSTRAP

Inferential procedures, such as a confidence interval estimate of a population mean, tend to be mysterious entities in introductory statistics, involving the substitution of some quantities derived from laborious calculations (like the standard deviation) into a magical formula that produces results very difficult to interpret. A simulation approach takes a lot of the mystery and magic out of this process and allows students to discover the notion of a confidence interval estimate without the use of formulas. This is becoming a widely used technique that probably needs only a passing mention here. A quick illustration can be seen by considering the following 25 measurements (considered to be a random sample) of the diameter of a tennis ball.

Diameters in millimeters
62 64 64 65 65 65 65 65 66 66 66 67 68
61 63 64 64 64 64 65 67 62 66 65 63

The sample mean and standard deviation here are 64.64 and 1.66, respectively. But, what is the "true" diameter? If these are random measurements they should surround the true value and so it is reasonable to take the true value as the mean of the conceptual population from which this sample came. To get anywhere in a simulation we need a model for this conceptual population, and a normal model is not unreasonable for these data. Now we can simulate what happens to sample means for samples of size 25 from normal distributions with various means and with the standard deviation about 1.66 (assuming this to be a good estimate of the variation in measurements). With the true means chosen to be 64, 65, and 66, the simulated distributions of possible values of the sample means (the so-called sampling distributions) are shown in figure 11.12. It is easy to see that a sample mean of a little under 65 is not going to come from a population with mean as low as 64 or as high as 66. The reasonable values for a population mean that could have produced this sample mean fall into a small interval (a confidence interval) around 65.

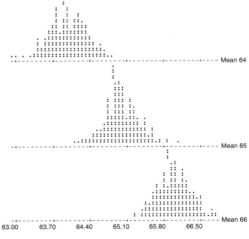

Fig. 11.12. Simulated sampling distributions of sample means

Modern resampling techniques, however, make the normal model unnecessary. A very bright statistician, Brad Efron (Efron and Tibshirani 1993), realized that if all we know about the population is what we see in the sample, then it is a good idea to assume that the population looks just like the sample except that it contains many more copies of these data values in the same relative proportions as seen in the sample. So, why not place the sample values for the diameters in a box and take samples of size 25 out of the box with replacement, calculating the observed sample mean each time? This should produce a distribution of possible values of the sample mean, much as was done from the normal distributions. (Of course, it is possible that some samples could have the same value repeated 25 times, but this is not very likely.) These are called bootstrap samples, and a resulting bootstrap distribution looks like the one in figure 11.13. Since there are 200 dots here representing 200 bootstrap samples, a reasonable way to arrive at a bootstrap interval estimate for the true population mean is find the middle 95 percent of the observed bootstrap means. If the smallest 5 and largest 5 of these means are lopped off, the resulting interval is (63.96, 65.20), a rather short interval around 65 and very close to what was produced by the normal models. Do not be misled, however: the bootstrap is not the same as the classical normal theory estimation procedure, but the results are often comparable when dealing with means.

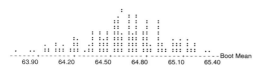

Fig. 11.13. A bootstrap distribution of sample means

There is nothing in a bootstrap procedure that ties it uniquely to means, and it can be used for virtually any statistic. For example, the areas of the states (in the first data set) have a very skewed distribution, and it would be better to take the population median rather than the mean as the measure of center. A random sample of 10 areas was selected, and they gave the following values.

Sample areas (thousands of square miles)
    53.2   9.6   42.1   35.4   52.4   83.6   147.0   114.0   268.6   77.1

Repeated bootstrap samples gave distribution of bootstrap medians shown in figure 11.14, for which the middle 95 percent yield an interval of (42.1, 130.5). Notice that this small sample results in a wide interval. It does, however, capture the population median of

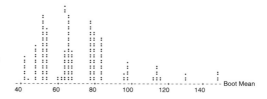

Fig. 11.14. A bootstrap distribution of sample medians

57.1. The bootstrap (Do you feel as though you have been pulling yourself up by your own bootstraps?) is a widely applicable technique that is simple and elegant in principle and easy to use with modern technology. It may well become a mainstay of introductory statistics courses in the foreseeable future.

DESIGNING STUDIES: SAMPLE SURVEYS AND EXPERIMENTS

The last topic to be considered perhaps should have been the first, because data should always be collected for a purpose and with a plan. In short, there are only two different types of designed studies that produce data meeting the assumptions required for statistical inference; they are sample surveys and experiments. Sample surveys require random samples from a fixed population and have as their goal the estimation of a population quantity (parameter). The example of estimating the median area of the states from a random sample of 10 states fits perfectly into this setting. Experiments begin with a fixed set of experimental units, such as patients in a clinic who have signed informed consent forms, white mice in a laboratory, or plots of wheat in a field. The goal of an experiment is to compare results from two or more treatments that are randomly assigned to these units. These two basic concepts of planning studies are quite distinct (random selection is not the same as random assignment) yet are often merged together in introductory statistics.

An excellent paper by Ludbrook and Dudley (1998) points out these distinctions and illustrates the analysis of a designed experiment by comparing hypothetical blood cholesterol concentrations between 7 men randomly assigned to eat fish and 5 men randomly assigned to eat meat for a period of time. The data are as follows.

Fish: 5.42 5.86 6.16 6.55 6.80 7.00 7.11   Meat: 6.51 7.56 7.61 7.84 11.50

The observed difference in means is 1.79. If these 12 numbers are put in a box and randomly divided into groups of 7 and 5 repeatedly, with the difference in means calculated each time, the observed difference will be exceeded with probability .0088. This randomization test thus leads to the rejection of the hypothesis that the mean difference in response for these two treatments could be zero. The treatments differ for these 12 men, but the generalizability of these results to a larger population of men is problematic and depends on how well these 12 represent a larger group. Experiments often do show different results for different subjects, and this problem of generalizing the results of an experiment is one of the main reasons why the public is confused about the seemingly conflicting results of experiments they read about in the media. It is a problem that should be dealt with early and often in the statistics education process.

An Overview of Important Statistical Concepts

The following outline shows a possible order for topics that would be consistent with modern statistical practice and would allow the topics to grow as a student moves through the grade levels. It emphasizes the structure present in data collection, data presentation, and data analysis. Examples are included in the list to illustrate the types of activities for students that are being suggested, but many such activities could be used at each stage. With regard to probability, a teacher should understand that statistics and probability are not the same thing. From a statistical perspective, probability should emphasize relative frequency interpretations and models for distributions of data; counting rules and details on the mathematics of probability would be better left to areas of discrete mathematics or calculus.

Elementary School
- A. Sort and count, with bar graphs (Sort buttons by color or style)
- B. Order and count, with bar graphs (Sort books by size)
- C. Measure, with histograms and stem-and-leaf plots (Measure height, length of foot, time to complete a task)
    - Measures of center—median and mean
- D. Census of class, school, or neighborhood (How many students have pets?)
    - Avoid sample surveys; the roles of randomness and inference come later.
- E. Relative frequency notion of probability
    - Empirical—What fraction of the time does the spinner land on red?
    - Theoretical—What do the relative areas suggest about the fraction of times the spinner should land on red?
- F. Simple experiments (Is the spinner random? How much does heart rate increase with exercise?)
- G. Association between variables
    - Categorical versus categorical (Do students who have siblings also tend to have pets?)
    - Measurement versus categorical (Do boys have fewer buttons on their clothes than girls have on theirs? Which books are used more, those about animals or those about people?)
    - Measurement versus measurement (How is height related to arm span? How does distance from home to school relate to travel time?)

Middle School
  A. Review all the above at more sophisticated levels of counting and measuring.
  B. Add numerical measures of spread (calculated with appropriate technology)
     • Quartiles (percentiles) and standard deviation, box plots
  C. Add an introduction to informal inference through simulation (using appropriate technology).
     • A test of significance for an experiment (Does a spinning penny land heads up half of the time?)
     • Margin of error for a sample survey with simple random sampling (What proportion of the students in the school eat breakfast regularly?)
  D. Add measures of association (without formal inference)
     • Difference between proportions for categorical versus categorical comparisons (Is the proportion of girls who eat breakfast regularly larger than that for boys?)
     • Difference between means for measurement versus categorical comparisons (Is the mean time for a hot liquid to cool to room temperature greater for a paper cup or a Styrofoam cup?)
     • Correlation and least squares regression line for measurement versus measurement comparisons; scatterplots (What is the relationship between arm span and height?)

High School
  A. Review all of the above at more sophisticated levels
  B. Add more on the design of experiments
     • Randomization to reduce bias
     • Comparative experiments
     • Blocking to reduce variation (Design an experiment to see which of two advertisements is more attractive to students.)
  C. Add more on the design of sample surveys
     • Randomization to measure margin of error
     • Measurement versus sampling bias
     • Stratification to reduce variation (Design a survey of neighborhoods in your city or town to estimate the proportion of voters favoring a local school issue.)
  D. Add more on inference
     • Tests of significance with p-values (introduced by simulation)
     • Estimation with confidence intervals (introduced by simulation)
  E. Add modeling (What is the pattern of growth in the number of autos on the road?)
  F. Add transformations of data

- Scales are chosen for convenience of analyzing and presenting information (pH for soil acidity and the Richter scale for earthquakes are both on a log scale)

Statistics is a process of thinking through a problem from inception, to clarification, to data, to analysis, to conclusion. This process involves other areas of mathematics (number concepts, geometry, algebra, functions), and the teaching of statistics should emphasize the interrelationships among various areas. In addition, statistics involves communication of ideas, both in understanding the practical problem that begins the process and in stating conclusions that others can understand. Teachers of mathematics should understand the process of statistical reasoning as well as the component parts of statistical methodology they may be called on to use. In fact, the process is more important than the parts.

CONCLUSION

Statistics has become and will remain an important component of grades K–12 education because its uses and misuses permeate the fabric of societies throughout the world. Statistics is built around data analysis, but this analysis requires careful thought throughout the processes of (1) exploring the data, (2) finding apparent associations among variables, (3) making inferences appropriate to the data and, in fact, (4) planning the data collection itself. With the use of appropriate technology, modern methods of data analysis can be introduced early so that students develop a clear and unified picture of statistics as a way of thinking about quantitative information.

REFERENCES

Cobb, George. "Teaching Statistics." In *Heeding the Call for Change,* edited by Lynn A. Steen, pp. 3–43. Washington, D.C.: Mathematical Association of America, 1992.

College Board. *Advanced Placement Course Description: Statistics.* New York: College Board, 1998.

Efron, Bradley, and Robert J. Tibshirani. *An Introduction to the Bootstrap.* New York: Chapman and Hall, 1993.

Hoaglin, David C., and David S. Moore, eds. *Perspectives on Contemporary Statistics.* Washington, D.C.: Mathematical Association of America, 1992.

Hoaglin, David. C., Frederick Mosteller, and John Tukey. *Understanding Robust and Exploratory Data Analysis.* New York: John Wiley, 1983.

Ludbrook, John, and Hugh Dudley. "Why Permutation Tests Are Superior to T and F Tests in Biomedical Research." *The American Statistician* 52 (May 1998): 127–32.

National Council of Teachers of Mathematics (NCTM). *Principles and Standards for School Mathematics.* Reston, Va.: National Council of Teachers of Mathematics, 2000.

12

Supporting Students' Ways of Reasoning about Data

Kay McClain

Paul Cobb

Koeno Gravemeijer

OUR purpose in this article is to describe how one group of students came to reason about data while developing statistical understandings related to exploratory data analysis. In doing so, we will present episodes taken from a seventh-grade classroom in which we conducted a twelve-week teaching experiment. (The first two authors shared teaching responsibilities during the teaching experiment.) One of the goals of the teaching experiment was to investigate ways to support middle school students' development of statistical reasoning proactively. Our interest was piqued by current debates about the role of statistics in school curricula (National Council of Teachers of Mathematics 1989, 1991; Shaughnessy 1992). The image that emerged for us as we read this literature was that of students engaging in instructional activities in which they both developed and critiqued data-based arguments.

As we worked to develop an instructional sequence, we viewed the use of computer tools as an integral aspect of statistical reasoning rather than as technological add-ons. As such, the two computer tools we designed were intended to support students' emerging mathematical notions while simultaneously providing them with tools for data analysis. In the twenty-first century, access to information will continue to be enhanced by new technologies. This fact highlights the importance of providing students with

---

The research reported was supported by the National Science Foundation under Grant No. DMS-9057141 and RED-9353587 and by the Office of Educational Research and Improvement under Grant No. R305A60007. The opinions expressed are solely those of the authors.

opportunities to develop and critique data-based arguments in situations that are facilitated by technologies. It is therefore imperative that learning opportunities of this type in which students can develop deep understandings of important statistical ideas become central aspects of school curricula.

In the following sections of this paper, we first outline the intent of the instructional sequence we used in the seventh-grade classroom and describe the role of the computer tools in this sequence. Against this background we then describe the classroom and present episodes intended to highlight students' development of sophisticated ways to reason about data.

INSTRUCTIONAL SEQUENCE

As we began to design the instructional sequence to be used in the seventh-grade classroom, we attempted to identify the big ideas in statistics. Our plan was to develop a single, coherent sequence and thus tie together the separate, loosely related topics that typically characterize middle school statistics curricula. In doing so, we came to focus on the notion of distribution. This enabled us to treat notions such as mean, mode, median, and frequency as well as others, such as "skewness" and "spread-outness," as characteristics of distributions. It also allowed us to view various conventional graphs such as histograms and box-and-whiskers plots as different ways of structuring distributions. Our instructional goal was therefore to support students' gradual development of a single, multifaceted notion—that of distribution—rather than a collection of topics to be taught as separate components of a curriculum unit. In formulating hypotheses about how the students might reason about distributions, one of our primary goals was that students would think about data sets as entities that have properties in their own right rather than as collections of points (Hancock, Kaput, and Goldsmith 1992; Konold et al. in press; Mokros and Russell 1995). We conjectured that if students did begin to think about data in this way, they could then investigate ways of structuring data sets that would help them identify trends and patterns.

As we began mapping out the instructional sequence, we were guided by the premise that the integration of computer tools was crucial in supporting our mathematical goals. Students would need efficient ways to organize, structure, describe, and compare large data sets. This could best be facilitated by the use of computer tools for data analysis. However, we tried to avoid creating tools for analysis that would offer either too much or too little support. This quandary is captured in the current debate about the role of technologies in supporting students' understandings of data and data analysis. This debate is often cast in terms of what has been defined as expressive and exploratory computer models (Doerr 1995). In one of these approaches—the expressive—students are expected to recreate conventional graphs with only an occasional nudging from the teacher. In the other approach—the exploratory—students work with computer software that presents a range of conventional graphs with the expectation that the students will develop mature mathematical understandings of their meanings as they use them. The approach that we took when designing computer-based tools for data analysis offers a middle ground between the two approaches. It introduces particular tools and ways of structuring data that are designed to fit with students' current ways of understanding while simultaneously building toward conventional graphs (Gravemeijer et al. in press).

The instructional sequence developed in the course of the seventh-grade teaching experiment involved two computer minitools. In the initial phase of the sequence, which lasted for almost six weeks, the students used the first minitool to explore sets of data. This minitool was explicitly designed for this instructional phase and provided a means for students to manipulate, order, partition, and otherwise organize small sets of data in a relatively routine way. Part of our rationale in designing this tool was to support students' ability to analyze data as opposed to simply "doing something with numbers" (McGatha, Cobb, and McClain 1998). When data were entered into the tool, each individual data value was shown as a bar, the length of which signified the numerical value of the data point (see fig. 12.1).

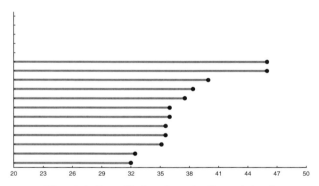

Fig. 12.1. Data displayed on the first minitool

A data set was therefore shown as a set of parallel bars of varying lengths that were aligned with an axis. Its use in the classroom made it possible for students to act on data in a relatively direct way. This would not have been possible had we used commercially available software packages for data analysis whose options typically include only a selection of conventional graphs. The first computer minitool also contained a value bar that could be dragged along the axis to partition data sets or to estimate the mean or to mark the median. In addition, there was a tool that could be used to determine the number of data points within a fixed range.

The second computer minitool can be viewed as an immediate successor of the first. As such, the endpoints of the bars that each signified a single data point in the first minitool were, in effect, collapsed down onto the axis so that a data set was now shown as collection of dots located on an axis (i.e., an axis plot as shown in fig. 12.2).

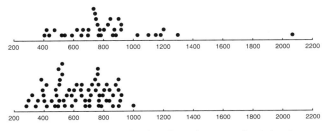

Fig. 12.2. Data displayed on the second minitool

The tool offered a range of ways to structure data. The tool's palate of options did not correspond to a variety of conventional graphs, unlike the palates of typical software packages that are available commercially. Instead, we designed the various options after identifying in the literature the various ways in which students structure data when they are given the opportunity to develop their own approaches while conducting genuine analysis (Hancock et al. 1992). Two of the options can be viewed as precursors to standard ways of structuring and inscribing data. These are organizing the data into four equal groups so that each group contains one-fourth of the data (precursor to the box-and-whiskers plot) and organizing data into groups of a fixed interval width so that each interval spans the same range on the axis (precursor to the histogram). However, the three other options available to students do not correspond to graphs typically taught in school. These involve structuring the data by (1) making your own groups, (2) partitioning the data into groups of a fixed size, and (3) partitioning the data into two groups of equal size. The first and least sophisticated of these options simply involved dragging one or more bars to chosen locations on the axis in order to partition the data set into groups of points. The number of points in each group was shown on the screen and adjusted automatically as the bars were dragged along the axis. The key point to note is that this tool was designed to fit with students' ways of reasoning while simultaneously taking important statistical ideas seriously.

As we worked to outline the sequence, we reasoned that students would need to encounter situations in which they had to develop arguments based on the reasons for which the data were generated. In this way, they would need to develop ways to analyze and describe the data in order to substantiate their recommendations. We anticipated that this would best be achieved by developing a sequence of instructional tasks that involved either describing a data set or analyzing two or more data sets in order to make a decision or a judgment. The students typically engaged in these types of tasks in order to make a recommendation to someone about a practical course of action that should be followed. An important aspect of the instructional sequence involved talking through the data creation process with the students. In situations where students did not actually collect the data themselves, we found it very important for them to think about the types of decisions that are made when collecting data in order to answer a question. The students typically made conjectures and offered suggestions about the information that would be needed in order to make a reasoned decision. Against this background, they discussed the steps that they might take to collect the data. These discussions proved critical in grounding the students' data analysis activity in the context of a recommendation that had real consequences.

Classroom Episodes

During the teaching experiment, the students explained and justified their reasoning in whole-class discussions. This was facilitated by the use of a computer projection system that allowed students to display the data sets and show the ways that they had structured the data. In addition, the students produced written arguments that we often used as a basis for classroom discussions. As the sequence progressed, we also asked the students to draw inscriptions in order to support their recommendations. (These inscriptions included nonconventional graphs and symbolic summaries of the data such as an axis marked with a three- or five-point summary.) In doing so they had to develop ways to show trends and patterns that supported their argument without reproducing data sets in their entirety.

One of the initial task situations that was posed to the students involved comparing two separate brands of batteries, Always Ready and Tough Cell. In discussing the task situation, students noted that a better battery would be one that lasted a long time. They framed the discussion around their experiences with using batteries in several types of electronic devices and reflected on how often they had to purchase batteries. They then offered numerous suggestions for how batteries might be tested. Against the background of this discussion, they were asked to analyze data from a sample of ten batteries of each of two brands that had been tested to determine how long they would last (see fig. 12.3).

The students then worked in pairs on the computers and used the minitool to organize and structure the data in order to help them make a decision. Afterwards, they discussed the results of their analysis in a whole-class setting.

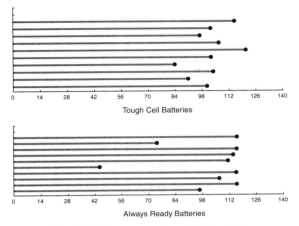

Fig. 12.3. Data on two brands of batteries

Celia was the first student to share her argument. She began by explaining that she used the range tool to identify the top ten batteries out of the twenty that were tested. In doing so, she found that seven of the ten longest lasting were Always Ready batteries. During the discussion of Celia's explanation, Bradley raised his hand to say that he compared the two brands of batteries a different way.

> *Bradley:* Can you put the representative value on 80? Now, see there's still [Always Ready batteries] behind 80, but all the Tough Cell is above 80 and I'd rather have a consistent battery that is going to give me above 80 hours instead of one. I just have to guess.
>
> *Teacher 1:* Questions for Bradley? Janine?
>
> *Janine:* Why wouldn't the Always Ready battery be consistent?
>
> *Bradley:* All your Tough Cells is above 80 but you still have two behind 80 in the Always Ready.
>
> *Janine:* Yeah, but that's only two out of ten.
>
> *Bradley:* Yeah, but they only did ten batteries and the two or three will add up. It will add up to more bad batteries and all that.
>
> *Janine:* Only wouldn't that happen with the Tough Cell batteries?
>
> *Bradley:* The Tough Cell batteries show on the chart that they are all over 80 so it seems to me they would all be better.
>
> *Janine:* (nods okay).

Bradley based his argument on the observation that all the Tough Cell batteries lasted at least 80 hours. He used the value bar to partition the data and determined that Tough Cell was a more consistent brand.

In comparing Celia's and Bradley's arguments, it is significant to note that while the students understood what Celia had done (compared the number of batteries of each brand in the top half of the data), her choice of the "top ten" was open to question. For instance, one student pointed out that if she had chosen the top fourteen batteries instead of the top ten, there would be seven of each brand. Celia's choice of the top ten was arbitrary in the sense that it was not grounded in the context of the investigation. Bradley, however, gave a rationale for choosing 80 hours that appeared to make sense to the students. He wanted batteries that he could be assured would last a minimum of 80 hours. As a consequence of this rationale for partitioning, his argument appeared to be accepted as valid.

The students continued to engage in similar investigations using the first computer minitool for several weeks. As they did so, many of them frequently used the value bar to partition the data. They would place the bar at a particular value along the axis and then reason about the number of data points that were above or below that value. It is important to note that the value at which they partitioned the data were typically not arbitrary. For instance, in a task about health care, many placed the bar at 65 years, arguing that this enabled them to focus on senior citizens in comparison to the rest of the population. We should clarify that we did not anticipate that the students would use the value bar in this way. Our expectation when we designed the tool was that they might use the bar to estimate the mean of a data set. Instead, the students adapted this feature of the minitool to their current ways of thinking about data.

As we worked with the students, we developed and modified tasks by analyzing their reasoning in each classroom session. We introduced the second minitool once we judged that the students had developed a variety of ways to organize the data and an understanding of what is involved in a data-based argument. Because the second tool displayed data as an axis plot, we were able to increase the number of data points in the sets.

One of the initial investigations with the second computer tool involved analyzing data on driving speeds on a very busy road in the city. As drivers are known to speed on this particular stretch of highway, the police department had decided to set up a speed trap to try to slow the traffic. Students discussed what information would be necessary to determine if the speed trap was effective. After much discussion on both issues of safety related to speed and on the specifics of this particular problem, students were asked to compare data on the speeds of sixty cars before the speed trap was set up with the speeds of sixty cars a month later to decide if the speed trap was effective (see fig. 12.4).

After students had analyzed the data using the computer minitool, we asked them to develop a written argument that could be submitted to the Chief of Police. In the subsequent whole-class discussion, students read their reports as part of their explanations. The first argument was presented by Janine.

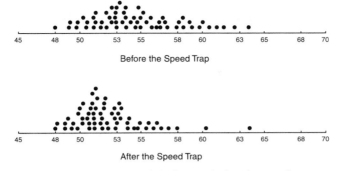

Fig. 12.4. Data on speeds before and after the speed trap

Janine: If you look at the graphs and look at them like "hills," then for the before group it is more spread out and more are over 55. If you look at the after graph, then more people are bunched up closer to the speed limit, which means that the majority of the people slowed down.

After Janine finished, the teacher used the projection system to display the data on the white board.

Teacher 1: Okay, Janine said if you look at this like hills ... now think about this as a hill *(draws a hill over the data in the first data set as shown in fig. 12.5)* and think about this as being a hill *(draws a hill over the second data set as shown in fig. 12.5).* See what Janine was talking about? Before the speed trap the hill was spread out, but after the speed trap the hill got more bunched up and less people were speeding.

Students seemed to understand Janine's argument and saw the relevance of her idea of "hills," as indicated by Kent's comment below.

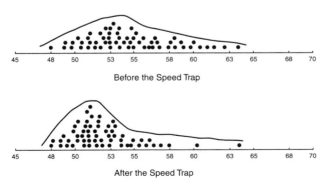

Fig. 12.5. Speed data with "hills" shown

Kent: They were slowing down. I want to compliment Janine on the way that she did that. I couldn't find out some way to compare, and I think that was a good way.

In this particular task, students attempted to find a way to organize and describe the data so that they could made a recommendation. The key point for us is that the notion of a data set as a distribution of data points emerged as the students discussed Janine's "hills" interpretation. She was concerned with how the data were distributed and focused on qualitative proportions of the data (e.g., the majority). In this episode, the students began to reason about global trends in entire data sets. Previously, they had focused on the number of data points in parts of data sets.

A further shift in the students' reasoning can be seen in an episode that occurred six days later. In developing the task, we had reasoned that we needed to find a situation in which the number of data points in the two data sets were very different. We hoped that this would lead to opportunities to question students' arguments that involved simple partitions and additive reasoning rather than proportional reasoning. The task we developed asked the students to analyze the T-cell counts of two groups of AIDS patients who had enrolled in different treatment protocols. A lengthy discussion revealed that the students were quite knowledgeable about AIDS and understood the importance of finding an effective treatment. Further, they clarified the relation between T-cell counts and a patients' overall health: increased T-cell counts are desirable. In the task, students were given data on the T-cell counts of 46 patients in a new, experimental treatment and the T-cell counts of 186 patients in a standard protocol (see fig. 12.6). The students were asked not only to make a recommendation about which protocol was more effective, but also to develop inscriptions that could be used to support their arguments.

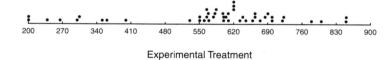
Experimental Treatment

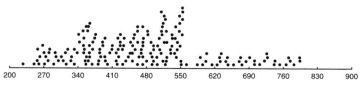
Traditional Treatment

Fig. 12.6. Data on T-cell counts of AIDS patients

As the students worked both at their computers and in groups, we monitored their activity in order to select students whose arguments might provide opportunities for shifts in mathematical thinking to occur. In one of the first reports that was discussed, the students had partitioned the data at a T-cell count of 550; they found that most of the data in the standard protocol was below a 550 T-cell count and most of the data in the experimental protocol was above a 550 T-cell count. During the discussion, the teacher clarified that these students had chosen the T-cell count of 550 because the "hill" of one data set was mostly below this value and the "hill" of the other was mostly above (see fig. 12.7).

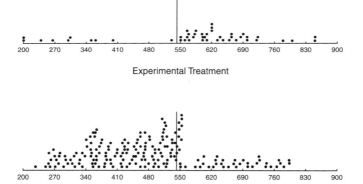

Fig. 12.7. AIDS data with partition between the "hills"

Thus, the students had partitioned at a particular value in order to develop a quantitative description of a perceived qualitative difference between the two data sets. Toward the end of the discussion, Janine made the following comment.

Janine: I think it would be helpful to know how many of the possible [patients] were in that range.

At a student's suggestion, the teacher then drew a diagram that recorded the number of patients in each treatment with T-cell counts above and below 550 (see fig. 12.8).

Teacher 2: Hey, I've got a question for everybody. Couldn't you just argue rather convincingly that the old treatment was better 'cause there were 56 people over 550? Fifty-six patients had T-cell counts greater than 550, and here there are only 37, so the old has just got to be better. I mean, there are 19 more in there, so that's the better one, surely.

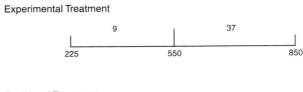

Fig. 12.8. Inscription showing partitions of AIDS data

Bradley: But there is more in the old.

Jake: Thirty-seven is more than half of 9 and 37, but 56 is not more than half of 130 and 56.

Kent: I've got a suggestion. I don't know how to do it *(inaudible).* Is there a way to make 130 and 156 compare to 9 and 37? I don't know how…

Kent's suggestion indicates that he wanted to find a way of comparing the information in the diagram. It was as if he was stating design specifications for the ideas of relative frequencies. The comments of the other students reveal that they also questioned the teacher's additive argument. The teacher capitalized on these contributions by introducing percentages as a solution to Kent's proposal. The students quickly calculated the percentages of patients in each treatment above and below a T-cell count of 550 and used the results to substantiate their initial arguments. After they had finished, Bradley made an observation.

Bradley: See, coming up with the percent of data cancels all the different numbers of data.

In the next report that was discussed, the students had used the computer minitool to organize the data into four equal groups. The inscription they developed consisted of axes marked only with each of the resulting intervals, similar to a box-and-whiskers plot (see fig. 12.9).

Bradley: On the four equal groups you can tell where the differences is in the groups.

Teacher 1: Can you do that by looking at this [inscription]? So, what do you see when you look at this, Bradley?

Bradley: That the new treatment was better than the old treatment.

Teacher 1: And what are you basing that comment on?

Bradley: Because the three lines for the equal groups [for the new treatment] were all above 525 compared to only one on the old one.

Experimental Treatment

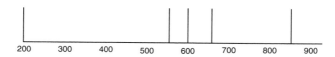

Traditional Treatment

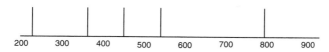

Fig. 12.9. Inscription of AIDS data partioned into four equal groups

In this exchange, Bradley clarified the conclusions that could be drawn from the inscription. He could see that 75 percent ("three lines") of the patients in the new treatment had T-cell counts greater than 525, whereas only 25 percent ("only one") of the patients in the old treatment had comparable T-cell counts. For him, this justified recommending the new treatment.

In the subsequent discussion, students focused on not only the validity of the argument but also the adequacy of the inscription. In doing so, they began to develop ways of using the four-equal-groups inscription to make arguments about the data. As they worked to understand the inscription, the validity of the argument was strengthened.

Martin: I think it would help to have the numbers to know how many were in each group.

Bradley: It doesn't really matter where all the data is, 'cause you know which group is better, 'cause you know where the data is. 'Cause you already see where the data is in the groups.

Other comments that Bradley made indicated that he (and several other students) had come to understand that 25 percent of the data were in each interval. He could therefore "see where the data is in the groups," whereas Martin needed to know exactly how many data points were in each interval. Differences such as these indicate the range in the students' interpretations. As a consequence of this diversity, the teacher encouraged the students to explain their reasoning by using the computer minitool and the inscriptions they had created. This enabled students such as Martin to contribute to the discussion of the four-equal-groups inscription in personally meaningful ways.

By the end of the classroom teaching experiment, over half of the students routinely described a part of a distribution as a proportion or percentage of the whole. In doing so, they reasoned about what Konold and colleagues (in press) have called *group propensities* (i.e., the rate of occurrence of some data value within a group that varies across a range of data values). Konold and colleagues argue that propensity is at the heart of what they call a *statistical perspective*. As we have seen, many of the students also structured data sets using what the statistician David Moore calls the *five-point summary* (i.e., extreme values, median, and quartiles) in order to characterize differences between distributions for a specific reason or purpose. In addition, their arguments now involved justifying the statistics they used to compare data sets. For example, students justified partitioning the data at a T-cell count of 550 based on the location of the hills as opposed to an arbitrary value such as the midpoint of the range. The whole-class discussions during the final sessions also indicate that almost all the students were now able to make and understand arguments that focused on patterns in how the data were distributed. It is significant to note that the students often changed their initial judgments in the course of whole-class discussions. This reveals that their ways of reasoning were constantly being challenged and modified by other's arguments.

Conclusion

We would stress that the purpose of the instructional sequence was not that the students might come to create specified graphs in particular situations or calculate measures of central tendency correctly. Most could already do the latter, although with little understanding. Instead, it was that they might develop relatively deep understandings of important statistical ideas as they used the computer minitools to structure data and make data-based arguments. It was for this reason that whole-class discussions throughout the classroom teaching experiment focused on the ways in which students organized data in order to develop arguments. In addition, students seemed to reconceptualize their understanding of what it means to know and do statistics as they compared and contrasted solutions. The crucial norm that became established was that of explaining and justifying solutions in the context of the problem being explored. This is a radically different approach to statistics than is typically introduced in middle schools. It highlights the importance of middle school curricula that allow students to engage in genuine problem solving that supports the development of central mathematical concepts.

There is much talk of preparing students for the information age but without fully acknowledging that the information in this new era will be largely statistical in nature. Cast in these terms, statistical literacy for the twenty-first century will involve reasoning with data in relatively sophisticated ways. New computer tools will provide opportunities for students to deal with information that is readily available, but often unmanageable with only paper and pencil. Classroom experiences in which students use computer-based tools to help them think and reason about problem situations serve to prepare them for the twenty-first century.

REFERENCES

Doerr, Helen. "An Integrated Approach to Mathematical Modeling: A Classroom Study." Paper presented at the Annual Meeting of the American Educational Research Association, San Francisco, April 1995.

Gravemeijer, Koeno, Paul Cobb, Janet Bowers, and Joy Whitenack. "Symbolizing, Modeling, and Instructional Design." In *Symbolizing and Communicating in Mathematics Classrooms: Perspectives on Discourse, Tools, and Instructional Design*, edited by Paul Cobb, Erna Yackel, and Kay McClain. Mahwah, N.J.: Lawrence Erlbaum Associates, in press.

Hancock, Chris, James Kaput, and Lynn Goldsmith. "Authentic Inquiry with Data: Critical Barriers to Classroom Implementation." *Educational Psychologist* 27 (1992): 337–64.

Konold, Cliff, Alexander Pollatsek, Arnold Well, and Allen Gagnon. "Students Analyzing Data: Research of Critical Barriers." *Journal for Research in Mathematics Education* (in press).

McGatha, Maggie B., Paul Cobb, and Kay McClain. "An Analysis of Students' Statistical Understandings." Paper presented at the Annual Meeting of the American Educational Research Association, San Diego, April 1998.

Mokros, Jan, and Susan Russell. "Children's Concepts of Average and Representativeness." *Journal for Research in Mathematics Education* 26 (January 1995): 20–39.

National Council of Teachers of Mathematics. *Curriculum and Evaluation Standards for School Mathematics.* Reston, Va.: National Council of Teachers of Mathematics, 1989.

———. *Professional Standards for Teaching Mathematics.* Reston, Va.: National Council of Teachers of Mathematics, 1991.

Shaughnessy, Michael. "Research on Probability and Statistics: Reflections and Directions." In *Handbook of Research on the Teaching and Learning of Mathematics*, edited by Douglas Grouws, pp. 465–94. New York: Macmillan, 1992.

13

Talking about Math Talk

Miriam Gamoran Sherin

Edith Prentice Mendez

David A. Louis

FOR the past several years, the authors of this paper have been investigating how teachers learn to build and support mathematical discourse communities. Such communities play a central role in improving students' understanding in the mathematics classroom. The premise, plainly speaking, is that the more students talk about mathematics, the more students learn about mathematics. Of course, discourse by itself does not necessarily improve students' learning. However, by explaining one's own ideas, evaluating other students' methods, and posing questions for the class to explore, students can develop deep understandings of mathematics (National Council of Teachers of Mathematics [NCTM] 1991). Students' talk about mathematics can also affect teachers' learning. More specifically, situations in which students explain their ideas and discuss their methods give teachers opportunities to rethink their own understandings of mathematics and to adjust their lessons to students' needs (Heaton 1994; Sherin 1996; Wood, Cobb, and Yackel 1991). Thus, as we seek to improve students' learning and help teachers make sustained changes in their practices, the development of discourse communities should become a central goal for mathematics education reform in the twenty-first century.

NCTM's *Professional Standards for Teaching Mathematics* (1991) envisions a broad range of classroom discourse—oral and written, whole-class and small-group, and discourse enhanced by technology. Deborah Ball and Magdalene Lampert, for example (Ball 1993; Lampert 1990), present examples of students making, supporting, and revising conjectures, and arguing for and against particular interpretations of mathematics. In another case, Brown, Campione, and their colleagues (Brown et al. 1991) find that providing students access to experts through e-mail promotes a worthwhile discourse between students and experts and enhances students' ability to do research in science. Likewise, Tom Banchoff and his students hold Web-based discussions about mathematics that broaden the meaning of "classroom" and encourage ongoing mathematical communication (1998). Although such examples help document why discourse is a valuable part of a rich mathematics classroom, an important question remains: How do teachers establish such classroom environments?

In this paper, we present a simple model for how teachers can establish discourse communities in their classrooms. To be sure, teachers already do many of the things employed by the model. However, what we are promoting is overt and repeated attention to the model as a whole. With David Louis as the teacher, we experimented with different ways to achieve a discourse community and documented both the teacher's and the students' roles in this process. Our research took place in one of Mr. Louis's eighth-grade mathematics classes at a middle school in the San Francisco Bay area.

What we have found is that whole-class discussion can be enhanced by the introduction of discourse strategies for students. Here we discuss three such strategies that were central to Mr. Louis's attempts to build a discourse community: (*a*) *explain,* (*b*) *build,* and (*c*) *go beyond.* These strategies are designed to help students do more than just talk—they are designed to help students engage in productive talk about mathematics. In brief, the *explain* strategy involves giving a reason for a particular idea or stating how you arrived at a specific result. *Build* refers to building on other students' ideas. *Go beyond* involves generalizing from a particular example to a broader mathematical issue. We believe that a crucial function of these strategies is that they provide students with explicit expectations for participation, thus increasing students' ability to engage in mathematical discourse. Moreover, because these strategies involve examining a diverse array of ideas, they invite participation from a wide range of students.

In addition to the students' strategies themselves, Mr. Louis's techniques for teaching students to employ the different strategies are noteworthy. Mr. Louis's method involves a number of components. He supports the students as they attempt to use a particular strategy, giving guidance and asking probing questions, and then fades into the background as his students' need for these supports diminishes. Mr. Louis also models using the strategy himself where appropriate. Furthermore, he conducts debriefings with the class where students explicitly reflect on the purpose of the various strategies and discuss particular ways in which the strategies have been used. We have found that these strategies are also useful as a means by which the teacher can assess the level of discourse in the classroom (Mendez 1998). For example, Mr. Louis evaluated particular discussions by considering whether and how students engaged in the three strategies outlined here.

STUDENTS' DISCOURSE STRATEGIES IN USE: EXAMPLES FROM THE CLASSROOM

We discuss the three student discourse strategies in the context of a lesson from Mr. Louis's classroom. The lesson, which took place during the first month of school, illustrates how each of the three strategies was integrated into classroom practice. On this day, the class was discussing a homework assignment on interpreting graphs. The assignment comes from a set of materials developed by the Joint Matriculation Board and the Shell Centre for Mathematics Education (1985). The figures presented here are our own reproductions of selected graphs from the assignment.

The students had been given sketches of six different containers and nine graphs. Each graph represented the height of the water level in a container as the amount of water in the container increased at a constant rate. Students were expected to match each container with its corresponding graph. For example, because the bucket fills up more and more slowly as water is added, the matching graph would show a curve with a decreasing slope (fig. 13.1).

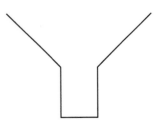

Fig. 13.1. A bucket and its associated graph

We join the discussion as Mr. Louis asked the class which graph best represented the changing height of a plugged funnel (fig. 13.2).

A number of students suggested that graph *b* corresponded to the plugged funnel (Fig. 13.3). They claimed that the linear portion of the graph corresponded to the bottom half of the plugged funnel, and that where the plugged funnel begins to widen, "the graph begins to curve." Other students argued that graph *c* represented the plugged funnel most accurately. These students seemed to focus on the distinction between the two linear segments of the graph. They explained that these two segments corresponded to the two segments of the plugged funnel. Furthermore, they iden-

Fig. 13.2. A plugged funnel

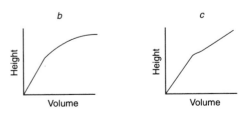

Fig. 13.3. Which graph best represents the plugged funnel?

tified a point on graph *c* where the two line segments met, and they perceived no such point on graph *b*. This point seemed to be a salient feature of graph *c* for many students.

Explain

As discussed above, the goal of the *explain* strategy is for students to justify their statements. In particular, Mr. Louis wanted students to understand that it was not enough simply to state an answer. Students also needed to give a reason for why their answer was reasonable or tell how they arrived at that conclusion.

One way that Mr. Louis encouraged students to give explanations was by using a simple prompt following a student's response. For example, when Tina suggests that graph *c* best represents the plugged funnel, Mr. Louis asks her to explain why. Tina then responds by making her reasoning overt for the class.

> Mr. Louis: You think it's what?
> Tina: c.
> Mr. Louis: Can you explain why?
> Tina: Because, there's a point. It's not like the bucket where it just goes like that *(motions a straight line with hand)*. There's a point, and so it's going like that *(motions two piecewise linear segments with hand)* instead of like that *(motions a straight line with hand)*.

At other times, Mr. Louis supported the students by pointing out where an explanation needs further clarification. Here, Ben explains why he thinks graph *b* represents the plugged funnel, and Mr. Louis asks Ben to elaborate on his description of the graph.

> Ben: I chose *b* because, when, at the beginning…it's kind of like a rectangle. And the shape isn't widening, or shortening, or closing in. So it would stay the same, like kind of a diagonal line for a period of time. But… then as it widens, it takes longer to fill up, so that's why it kind of curves.
> Mr. Louis: How does it curve? Saying "it curves" isn't very descriptive. So, how does it curve?

While such prompts and questions were an important part of Mr. Louis's techniques for establishing the importance of explaining one's ideas, they were not always necessary. In many instances, students' initial responses included a justification. For example, in the following excerpt, Robert explains why he disagrees with Tina's claim.

> Robert: I disagree because when it starts to get wider at the top, it doesn't go straight to really wide like on the graph. It gradually gets wider and wider, so it would be curved.

Although it is still early in the school year, it seems possible that the expectation for explanation is already firmly established in the minds of some students.

Build

The notion that students would build on one another's ideas was central to Mr. Louis's vision of how his class would function as a discourse community. In his journal Mr. Louis wrote, "I realized that my students could not become active learners, nor could they become a mathematical community, if they did not first know how to engage in meaningful discussions with each other and believe that each of them had important knowledge to contribute to the class." Thus, Mr. Louis's goal was for students to listen and respond to one another's ideas and to recognize that students' ideas were a valued part of classroom discussion.

With simple questions, Mr. Louis often encouraged students to build on one another's comments. For example, after Tina explains that she believes that graph *c* best represents the plugged funnel because it shows a point where the graph makes a distinct change, Mr. Louis queries the rest of the class.

Mr. Louis: OK. So what do people think about that?

Mr. Louis's question to the class brought a string of responses. In fact, students quickly seemed to become comfortable with this type of question and often responded using the "I agree" or "I disagree" language. For instance, Robert, whose response is given above, starts by saying, "I disagree because…" The form of Amy's reply is similar, though she agrees with Tina.

Amy: I agree with Tina because…it stays parallel for a little while, and then it opens really wide. So I think that's a big jump.

In some cases, as students build on one another's comments they offered new information about the topic. Here, Jason responds to Tina's claim and gives additional justification for Tina's idea. Following this, Lia adds to the conversation by suggesting that an alternative container, one with a "wider rectangle" at the top, would match the description Tina had proposed.

Jason: I agree with Tina because there's a point in the funnel where it stops. It changes immediately, it doesn't change right away, I mean it doesn't change gradually. So there'd have to be a point in the graph, and that's the only graph that has one point in it. So that's the only choice.

Mr. Louis: OK. What do other people think? Anyone else? Lia?

Lia: Disagree with them because … it doesn't just suddenly, it's going at a steady pace. So then there would probably be a wider rectangle or something. But this is just gradually getting wider. It's getting wider, so then when you pour in the water, it goes a little slower than before but it's still faster than when it ends.

Another technique that Mr. Louis used to encourage students to integrate their comments was to rephrase and compare differing ideas that students had brought to the floor. This is akin to revoicing (O'Connor and Michaels 1996), but it contrasts the ideas of two students rather than rephrasing just one. For example, at one point in the discussion, Mr. Louis restates the two positions that are under debate.

Mr. Louis: I think what Robert's idea is that it would get wider gradually and Tina's idea is that it changes and goes straight because there's definitely a changing point.

By encouraging students to enter the discussion with their own opinions, Mr. Louis helped them build on one another's work and add to the ideas that had already been presented. Furthermore, such techniques proved quite successful. The students picked up on Mr. Louis's phrasing and used the "I agree with" language readily.

Go Beyond

The third discourse strategy we describe is called *go beyond*. The idea is that students would not only respond with an answer to a specific example or problem, but would also be able to generalize somewhat, to go beyond the specifics of the current case. We see Lia's response to Tina, presented in the preceding section, as a first step in this direction. Lia moves beyond the set of containers presented in the homework problem and describes an alternative container, one that would match Tina's claim.

In general, the *go beyond* strategy was not easily employed by students. However, Mr. Louis modeled this strategy in class, and it became a part of the discourse structure. Furthermore, the technique served dual roles. It enabled Mr. Louis to demonstrate the process of synthesis for the students as well as to introduce specific mathematical content into the discussion.

This strategy is evident in the continuation of the plugged-funnel discussion. After the students presented and discussed various arguments for graphs *b* and *c*, Mr. Louis introduced a new concept to help the students recognize that graph *b* more accurately represented the plugged funnel. He said, "Here's the concept that I think is making people think a little bit differently—constant slope." Mr. Louis then sketched a graph with a constant slope and a graph with a changing slope and discussed the meaning of the different slopes with the class (see fig. 13.4).

Using this information, the students resumed their discussion of how best to represent the plugged funnel. The class concluded that the upper half of a graph of the plugged funnel required a changing slope rather than a constant slope. In the following excerpt, Jason reiterates this finding using the new terminology. First he clarifies that, although somewhat difficult to see, there are two parts to graph *b*, similar to the two parts of the plugged funnel. The first part has a constant slope and the second part has a decreasing slope. In addition, he points out that the top half of graph *c* is constant, unlike the top half of the plugged funnel, whose width increases.

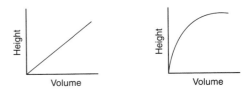

Fig. 13.4. Constant and changing slopes

> Jason: Well, it's hard to tell on if on b is there a point on the graph. Is that a point? Because it would be b if there was. The first strip was constant. And then there's like a point about halfway, and then it decreases. That would be it.
>
> Mr. Louis: Yeah, that's a good point.
>
> Jason: It can't be, it can't be c then, because the second half is a constant, right? And the funnel isn't constant.

Here, Jason begins to move past the specific task of matching a graph and a container to the more abstract problem of interpreting the features of a graph. Moreover, his focus on the slope of a graph is a widely applicable technique that can be used to examine the relationship between two changing quantities. While it was Louis who introduced this terminology and gave the discussion a specific mathematical focus, the students remained engaged. And together, the class was able to "go beyond" by generalizing from a particular problem to a broader mathematical issue.

Conclusions

Because mathematical discourse has the potential to affect students' learning and to provide opportunities for teachers to learn, it remains a central goal of reform efforts. Here we sought to describe three strategies that can aid in the development of a discourse community and that clearly support the learning of mathematics. To introduce these discourse strategies, we chose to draw from a single classroom discussion. We hope that providing detailed images of one teacher's practice will support both teachers and researchers in their efforts to develop and maintain classroom communities in which discourse is emphasized and students' ideas are valued.

Discourse has been recognized as an important component of learning since the time of Plato. Yet it takes on new meaning as we enter the twenty-first century. In particular, we expect that further developments in technology in the twenty-first century will make it increasingly possible to extend discourse communities outside the classroom. In such contexts, the use of *explain, build,* and *go beyond* may look somewhat different from what was presented in the example from Mr. Louis's classroom, but the strategies could still prove to be equally effective. For instance, the loss of face-to-face formats could lead to a loss of personal, spontaneous interactions between students and teachers. Teachers might find it a difficult task to nurture the trust, self-confidence, and mutual respect needed to sustain a discourse community and to get every student to participate constructively in the discourse. However, we believe that the three, simple-to-understand steps we have outlined in this article can help students learn how to engage in constructive discourse even when not face-to-face. Again, it is a matter of communicating and modeling over and over again what is expected of them in simple and clear terms.

In addition, the *explain, build,* and *go beyond* strategies might even be enhanced within such extended communities of learners. For example, students may be asked to post their solution strategies on the Web or to describe ideas in e-mail messages. Doing so would give students with increased time to articulate their ideas and may enhance students' ability to explain. In addition, building on one another's ideas may become a natural process as students respond to one another's ideas in e-mail. Similarly, an electronic discussion forum can be structured so that it displays who initiated a particular topic and what the various responses to the initial posting were. Such a record extends the meaning of *build* by keeping an explicit account of this process. Furthermore, as discussions extend over days, weeks, or months, additional opportunities can be provided for students to go beyond the specifics of a given problem and consider more general mathematical issues and ideas. Through such adaptations, the discourse strategies presented here will continue to be an important vehicle through which students learn to engage in productive discussions about mathematics.

REFERENCES

Ball, Deborah Loewenberg. "With an Eye on the Mathematical Horizon: Dilemmas of Teaching Elementary School Mathematics." *Elementary School Journal* 93 (March 1993): 373–97.

Banchoff, Thomas. "Internet-Based Geometry Course." Presented at the Carnegie Institute for the Advancement of Teaching, Menlo Park, Calif., February 1998.

Brown, Ann L., Doris Ash, Martha Rutherford, Kathryn Nakagawa, Ann Gordon, and Joseph C. Campione. "Distributed Expertise in the Classroom." In *Distributed Cognitions: Psychological and Educational Considerations,* edited by Gavriel Salomon, pp. 188–228. New York: Cambridge University Press, 1993.

Heaton, Ruth M. "Creating and Studying a Practice of Teaching Elementary Mathematics for Understanding." Doctoral diss., Michigan State University, 1994.

Joint Matriculation Board and Shell Centre for Mathematics Education. *The Language of Functions and Graphs: An Examination Module for Secondary Schools.* Nottingham, U.K.: Joint Matriculation Board and Shell Centre for Mathematics Education, 1985.

Lampert, Magdalene. "When the Problem Is Not the Question and the Solution Is Not the Answer: Mathematical Knowing and Teaching." *American Educational Research Journal* 27 (Spring 1990): 29–63.

Mendez, Edith Prentice. "Robust Mathematical Discussion." Doctoral diss., Stanford University, 1998.

National Council of Teachers of Mathematics. *Professional Standards for Teaching Mathematics.* Reston, Va.: National Council of Teachers of Mathematics, 1991.

O'Connor, Mary Catherine, and Sarah Michaels. "Shifting Participant Frameworks: Orchestrating Thinking Practices in Group Discussions." In *Discourse, Learning, and Schooling,* edited by Deborah Hicks, pp. 63–103. New York: Cambridge University Press, 1996.

Sherin, Miriam Gamoran. "Novel Student Behavior as a Trigger for Change in Teachers' Content Knowledge." Paper presented at the Annual Meeting of the American Educational Research Association, New York, April 1996.

Wood, Terry, Paul Cobb, and Erna Yackel. "Change in Teaching Mathematics: A Case Study." *American Educational Research Journal* 28 (Fall 1991): 587–616.

14

Blending the Best of the Twentieth Century to Achieve a Mathematics Equity Pedagogy in the Twenty-first Century

Karen C. Fuson

Yolanda De La Cruz

Stephen T. Smith

Ana María Lo Cicero

Kristin Hudson

Pilar Ron

Rebecca Steeby

As we enter the twenty-first century we carry some unsolved problems with us. Many children from all backgrounds do not understand mathematics enough to use it or cannot even do many tasks accurately. This problem is

---

The Children's Math Worlds research reported in this article was supported in part by the National Science Foundation under Grant No. RED935373 and in part by the Spencer Foundation. The opinions expressed in this article are those of the authors and do not necessarily reflect the views of NSF or of the Spencer Foundation. This article reflects the extraordinary support of the teachers with whom we worked and at least some of what we learned from the families and children with whom we worked.

Correspondence should be addressed to Karen C. Fuson, School of Education and Social Policy, Northwestern University, 2115 N. Campus Drive, Evanston, Illinois 60208-2610; office phone (847) 491-3794; e-mail fuson@nwu.edu.

especially acute for children of poverty or from homes where standard English is not spoken or is not the main language (e.g., Secada 1992). Many children leave third grade hopelessly behind, especially in urban schools. We share in this article an approach to teaching and learning that was developed in urban first- and second-grade classrooms in English and in Spanish. This approach did result in much higher levels of performance and understanding by our urban children (Fuson, Smith, and Lo Cicero 1997; Fuson 1996). This approach has more recently been extended successfully to third grade and to middle- and high-income children. We highlight some of the less familiar aspects of our pedagogy in the hope that it can help us enter the new century with approaches to overcome these problems. Some modifications are necessary for children older than grade 3, especially some source of extra support for homework, but the major elements apply to all grade levels.

Overview

We combined aspects of traditional pedagogical approaches with the most powerful elements of reformed teaching based on meaning-making to create our Toward a Mathematics Equity Pedagogy (TMEP) program. It uses the research on children's thinking as a basis for designing learning trajectories (Simon 1995) of increasingly advanced solution methods and then helps children move through these learning trajectories. We have on one page (see Table 14.1) summarized the attributes that we found to be effective in helping our children from urban backgrounds learn high levels of mathematics. The top three categories in the table focus on the reform aspects, and the bottom three summarize more traditional elements.

TABLE 14.1
Six aspects of TMEP

I. **Start where children are and keep learning meaningful. Use many meaning-focused classroom activities:**

<u>Referential classroom and math modeling: Mathematical words and symbols linked to meaningful referents</u>

The teacher leads mathematizing (seeing math in children's worlds) and rich language use by all.

Everyone's use of Meaningful Drawn Models (MDMs) facilitates the reflection, discussion, and analysis of everyone's thinking; MDMing enables teachers to analyze children's thinking and children's errors.

All children produce their own MDMs, language, questions, and problems (not just answers).

<u>Connected teaching: Cocreation of an inclusive and participatory classroom culture</u>

Table 14.1 (cont.)

    Class coconstructs emerging related understandings and builds on individual and shared MDMs.

    Curriculum and teacher connect math topics to children's lives and to their imaginations

    Mathematizing, MDMing, and rich language use

        a) make problems accessible to all by providing multiple levels of access (everyone can participate),

        b) validate and use all children's language and experiences while connecting them to standard language and symbols,

        c) facilitate listening, speaking, writing, and helping competencies.

    The teacher establishes coherence within a class, through the year, and across subject-matter areas.

II. Set high-level mathematical goals and expectations for all. Bring children up to the higher mathematics.

    Use a few ambitious core grade-level topics and cumulative experiencing and peer helping throughout the year.

    Construct ladders to mathematical concepts, symbols, methods, thinking, and discourse through

        a) meaning-focused classroom activities that help children build knowledge skills,

        b) work on prerequisite competencies to bring all children to mastery,

        c) activities to help children move through developmental progressions to more efficient and general methods.

    Teachers use assessment results to adapt their teaching and to help children focus their learning.

III. Develop a collaborative math talk culture of understanding, explaining, and helping.

    Connected knowing: Learning occurs within a supportive environment in which help is available from peers and from the teacher; emotional as well as cognitive needs are addressed.

    Separate knowing: All listeners can separate an idea from the person having the idea, so an idea can be respectfully modified or improved, or errors fixed, without diminishing its originator.

    All teachers are learners, and all learners (students) are teachers of themselves and of others (peer helping).

    Everyone expects that each child will understand his or her mathematical actions and use of MDMs and with practice will come to verbalize these understandings.

    Math talk connects to referents (e.g., MDMs, story situations) so that participants and listeners can understand.

    Debugging errors helps everyone learn more deeply; everyone including the teacher sometimes makes errors in math; no laughing or making fun about errors; errors deserve respectful and sensitive help from everyone.

Table 14.1 (cont.)

IV. Build on the best of traditional instruction.
> Teacher and children demonstrate and explain.
>
> Teacher assists children in learning productive roles in each classroom activity structure.
>
> Children practice and memorize after repetitive meaning-making experiences for concept building.
>
> Individuals solve problems with help available from a peer or teacher.
>
> Children do worksheets. (Ours facilitate the use of MDMs related to standard mathematical symbols.)
>
> Homework is somewhat similar to parental experiences. Mobilize a home helper to support homework.

V. Facilitate the learning of general school competencies. Increase self-regulatory actions so that children:
> Become more organized, including learning to do work neatly and regularly returning completed homework;
>
> Understand when they do not understand, and seek appropriate help when they need it;
>
> Are involved in setting some learning goals, making a plan for meeting them and carrying out the plan;
>
> Are helped to reflect on their own progress to affirm how much they have learned.

VI. Mobilize learning help in the home. Actualize home-school links focused on students' learning by:
> Designating a home helper to monitor and help with homework (daily homework is expected; much of the homework is familiar to parents, and most is similar to that experienced by the student in the classroom);
>
> Giving home games and activities to learn and practice prerequisite competencies and knowledge skills;
>
> Conducting home-helper learning sessions to improve their helping and understanding of new aspects of the curriculum.

Equity, to us, means a classroom in which each child is included and affirmed as an individual and in which access to mathematical competencies valued by the culture is provided to all children. Given that educational resources are always finite, equity means balancing the needs of various individuals and trying to organize socially to maximize the learning of all. Central to equity is setting high-level goals for all and then using various methods of learning support so that individual children can go as far as they can. In an analysis of our urban school experience (Fuson et al. 1999), we summarize how the frequent reduction of goals to accommodate the realities of some or many underprepared children prevent all children in urban schools from reaching high-level goals. Such a reduction of goals is common in bilingual settings (Moll 1992).

TMEP sets high-level goals for learning with understanding, for high-level oral language competencies, and for the sophisticated use of mathematical modeling and mathematical symbolizing. These high-level goals are achieved by enabling all children to enter the mathematical activity at their own level. Teachers accomplish this by using rich and varied language about a given problem so that all children come to understand the problem situation, by mathematizing (focusing on the mathematical features of) a situation to which all children can relate (and that may be generated by a child), and by having children draw models of the problem situation. Cumulative experiencing and practicing of important knowledge skills helps children move through developmental trajectories to more advanced methods. Peer helping provides targeted assistance when necessary. The knowledge of the helper also increases. Assessment provides feedback to all and permits realistic adjustments of proximal learning goals by children and by the teacher.

We found that we needed to consider affective, social, motivational, self-regulatory, and self-image aspects of learning and not just focus on building mathematical conceptions. Thus, children need to be helped to see themselves as included in the world of mathematics. They need to be taken seriously as learners so that they can begin to see themselves as learners. Mathematization (starting with children's experiences), the coconstruction of understandings in a collaborative classroom, and involvement in learning-goal setting and evaluation all contribute to students' growth. Children and their home experiences and families are valued and affirmed by their inclusion in word problems stemming from mathematics storytelling. We emphasize effort rather than "ability," which is often misdiagnosed in schools, and we have created many examples of what Resnick (1995) discussed as ability created by effort.

Traditional and reform practices are usually posed as alternatives. We found many elements of value in both perspectives, especially in the context of urban schools. We wove these elements into a fabric of teaching-learning activities that would support teachers' and children's construction and use of mathematical meanings linked to the traditional mathematical symbols and words of the culture. In each mathematics domain the teacher connects children's experiences, words, meanings, object and drawn representations, and methods to the traditional mathematical symbols, words, and methods of that domain. These meaning connections define what we call the referential classroom: referents for mathematical symbols and words are used pervasively within the classroom. The referential classroom provides the links between what Vygotsky called spontaneous concepts and scientific concepts (Vygotsky 1962, 1986). The attributes of such a referential classroom and their relationships to Vygotsky's theory are discussed in more detail in Fuson, Lo Cicero, et al. (1999) and Hiebert et al. (1997).

We do not have space here to discuss all aspects of TMEP. We will instead provide an overview of the least familiar but core aspects of our approach.

AFFECTIVE SUPPORTS

Our affective supports are described by the terms *connected teaching, connected knowing,* and *separate knowing* (from Belenky, Clinchy, Goldberger, and Tarule 1986). By connected teaching, we mean helping children in a classroom cocreate a culture within which each child can be affirmed, can steadily grow in a range of social and conceptual competencies, and can teach (assist others) as well as learn. Coherence is a key to emotional, social, and mathematical growth because urban young children frequently experience too little coherence in their lives. Our focus on the conduct of the whole child is consistent with the Latino broader meaning of *educación* (education) as *formación* (formation): forming the child to know proper ways to behave in many different settings. Our focus on active listening and helping stems from the Latino experience in family-centered, rather than individually-focused, living. Sharing and helping is a pervasive fact of everyday life for some children, thus making it relatively easy for them to function in these ways in a TMEP classroom. For children who have had less such experience at home, the focus on listening and helping in our classrooms provides a valuable learning opportunity.

FOUR PHASES BUILDING TO HIGH-LEVEL MATHEMATICAL GOALS

In the initial phase, discussion and activities begin where the children are. Teachers find out and build on what children already know and understand. In the second phase, teachers introduce activities that emphasize and support the learning of the mathematical structure of a given mathematics domain. If the domain is initially accessible to most children, as with many word-problem situations involving single-digit numbers, this phase merges with the first phase. Activities elicit individual children's own views of the structure through engaging children in making their own meaningful drawn models (MDMs) of the situation. If the domain has a lot of specialized cultural knowledge, such as multidigit numeration in words and in written symbols, the activities initially provide structured conceptual experience with this cultural mathematical knowledge so that children can learn these structures and the meanings of the symbols and words.

The third phase is the longest. Cumulative experiencing (cf. repetitive experiencing, Cooper 1991) of the mathematical structure enables children to construct robust conceptions of those structures that can be generalized, abbreviated, internalized, and used flexibly in problem solving. For many major mathematical concepts, this generalization, abbreviation, and internalization follows a developmental progression of identifiable levels through which children need to be supported. A great deal has been learned about these developmental progressions in the past 20 years, especially for the addition and subtraction of small whole numbers (see Fuson 1992a, 1992b for summaries of this literature and the chapters in Leinhardt, Putnam, and Hattrup 1992 for summaries of the literature in other areas). Teachers' understandings of these learning trajectories (Simon 1995) enable them to recognize where children are and then to help children move through these learning trajectories to more advanced and efficient methods. Helping children move through these progressions may take a long time—weeks and even months for some concepts and some children. Cumulative mathematical experiencing at each of these developmental levels is required.

The final learning phase has many characteristics of traditional drill and practice, although we think of it as consolidation because of the frequent connotation of practice as a rote activity. Practice is necessary to bring certain prerequisite understandings to a rapid and accurate level of competence. However, such final-phase practice must rest on a base of meaning that can be accessed during problem solving and used in the construction of higher-level concepts. For example, learning to count by tens is not just a rote, meaningless activity; it can be surrounded with various kinds of meaning supports (Fuson, Smith, and Lo Cicero 1997; Fuson and Smith 1995, 1997). But after the introduction of these meaning supports, counting by tens needs to be practiced in various ways until everyone in the class can count by tens rapidly and accurately. Multiplication and division concepts need to be introduced meaningfully, but later specific facts also need to be practiced so that they can be used in multidigit computation.

In TMEP, such meaningful practice is the culmination of extensive earlier activities in the first three learning phases. It is conceptually-based drill and practice. We call the outcomes of all these learning phases *knowledge skills*—skills based on knowledge—to distinguish them from the rote skills that are the goals of much of traditional mathematics teaching. Knowledge skills are rapid and flexible, but they rest on a system of knowledge that can be used in deployment of the skill. This helps children use these knowledge skills in different situations. An important aspect of this final phase is that teachers attempt to structure such practice to facilitate children's learning of self-regulatory actions. Children grow in monitoring their own progress and in exercising some choice in what they attempt. For example, third graders choose which count-bys (7, 14, 21, 28, and so on) and multiplication facts they work on using feedback from checkups.

Mathematizing and Rich Language Use

The core of the initial learning phase is mathematizing children's stories of their experiences and using rich language about these experiences (Hiebert et al. 1997; Lo Cicero, De La Cruz, and Fuson 1999; Lo Cicero, Fuson, and Allexsaht-Snider 1999; Ron 1999). The teacher elicits stories and helps children focus on, or extends the story to, potential mathematical elements that are the focus of that lesson. Children pose questions about the situation. They retell the situation in their own words. Children then pose mathematics problems from the situation. They frequently solve problems, and some children describe their solution methods. All this mathematizing and rich language use is done as a coconstructed process by the class as a whole orchestrated by the teacher as leader. The story is initially that of one child, but the participation of everyone as a contributor or as an active listener turns it into a shared class experience. The process attends to affective, social, and cognitive needs. Immigrant children can be connected to both their new and their old homes by telling stories about their grandparents or other family members they left in Mexico or elsewhere as well as by telling stories about their lives here.

Some stories are revisited throughout the year to provide continuity, coherence, and shared coconstructed referents. Everyone gains additional extended family feelings as they experience everyone's stories about their own lives and families. For example, one year a child told a story about a dog he had to leave with his grandfather in Mexico, and the teacher mathematized this story to focus on how many bones the dog ate. Related problems for the class to make up as the year went on concerned how far the dog ran, how many packages of food were needed for specified times if the food was packaged in boxes of ten, and how much food cost when it was purchased versus being leftovers. In other years repeated story scenarios concerned candy made and sold by one child's mother and apples grown on a farm.

Using Meaningful Drawn Models (MDMs)

The core of the second and third phases is extensive use of children's MDMs progressively linked to standard mathematical symbols and words. For example, children make drawings of a word-problem situation using circles and any other means of their choice to portray the mathematical relations in the situation (see fig. 14.1). These later become related to equations and to tables. This route permits early algebratizing of children's problem solving (Fuson, Hudson, and Ron in press). Children draw quantities to add and subtract multidigit numbers (see fig. 14.2 and Fuson, Smith, and Lo Cicero 1997; and Fuson and Smith 1995 for more details). These quantities are linked to, and over time are replaced by, mental and written numeral methods. Using such models resolves many management problems created by extensive use of manipulated objects as well as supporting reflection and discussion.

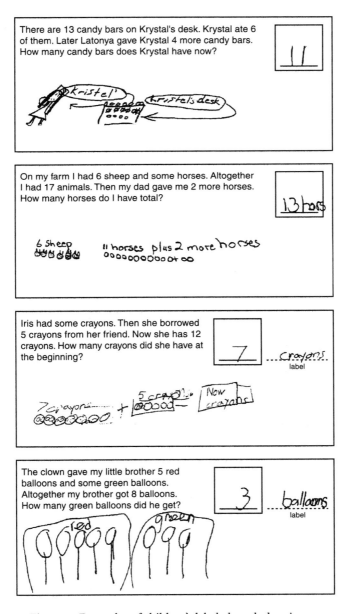

Fig. 14.1. Examples of children's labeled math drawings

MDMs give children experience in mathematizing situations to show their own conception of the mathematical structure. Children symbolize in different ways, so they can become aware of the goal of symbolizing in mathematics as well as see different kinds of symbolizing. This range of experiences with two-dimensional symbols drawn on paper is especially important for those children from low-literacy families who have comparatively little experience with symbolizing. Furthermore, MDMs leave records of thinking, thus facilitating reflection by their drawer and assessment of the drawer's thinking. Therefore MDMs facilitate the reform discussion styles of math talk—describing and justifying solution methods—by providing visual referents for such discussions. This enables everyone to look at the explainer's MDM during the discussion. Discussions that are merely verbal can lose many lower-achieving children (Murphy 1997). MDMs also allow the teacher to look at and think about a child's work outside class when there is time to reflect on and analyze it. Manipulatives leave no such records, and it is difficult for a teacher to get around to everyone during class to see each person's thinking.

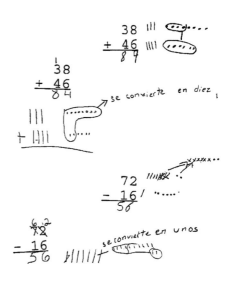

Fig. 14.2. Children's "ten-stick and dots" solution methods for two-digit addition and subtraction

Language Issues

The issues of language learning and language use are particularly important for Latino children, many of whom face schooling in a language that is not their native language and may not be spoken at home. We developed teaching-learning activities that directly support language learning and rich language use. These result in rich learning for all children, including the upper-middle-class children with whom some of our teachers have been working more recently. TMEP stimulates children to produce their own words and spatial drawn models of the mathematical aspects of a situation. Teachers elicit multiple ways to discuss and view the same math situation. Children make up problems and pose various questions. Children describe their models and their solution methods in a variety of representations.

A typical school approach to students who do not have strong English-language skills is to view mathematics as consisting solely of mathematical symbols and to reduce both the levels of mathematics taught and the use of language in such teaching and learning (Moll 1992; Secada 1992). We have found it more productive to increase the amount of language use, but to situate this use in referential situations that can provide active meanings for such language. For example, in the situations from children's lives described earlier that were returned to during the year, children would generate story problems of varied types using that situation. Other children would ask the mathematical question in varied ways to practice producing mathematical language in a context.

WORKING WITH FAMILIES

We found it to be crucial to involve families in their children's mathematical education. An overview of our project work with families is presented in De La Cruz (1999). We have concentrated on efforts that will help families support their child's daily learning in the mathematics classroom, their practicing of needed skills at home in family games, and their experiencing and noticing mathematics in the real world outside of school. We found that when efforts were made to establish home-school connections, human resources can be found within the homes of almost all children to support their mathematics learning. Teachers found it to be important to appreciate and affirm all of the efforts parents were making in raising their children. The building of home-school links also needs to be viewed by all participants as informationsharing, as building mutual adaptations between the school and home settings, and as involving joint working toward the common goal—mathematical learning by the family's child.

To achieve the high level of mathematics understanding and skill that was our goal, children needed to complete daily homework. Our urban children and families were very supportive of such homework. It made the children feel grown-up, and most of them enjoyed doing the homework. Families were involved by identifying a "math helper" in each home to be responsible for monitoring the child's homework completion and to help if necessary. Identifying a particular person for this responsibility was helpful in homes where many pressures create difficult ongoing and changing life demands. When families understood that it was important to do so, almost all would organize themselves to identify such a helper. This might be a parent, an older sibling, an uncle or grandparent who lived with the family or nearby, or a neighbor. Because of language differences, no phone in the home, and distance from the school, teachers had to be resourceful and persistent in order to communicate the need for a math helper to some families. Older siblings or children who lived nearby sometimes carried messages or communicated the need to meet with the teacher and also provided translating help at any meetings.

Most homework involved mathematical situations that placed few demands on family helpers. New elements of TMEP (e.g., the use of MDMs) were explained in notes sent home, at parent nights, and by the children themselves. Wherever possible, the homework was designed to be not too different from homework the home helper might have had. In some reformed mathematics curricula, parents are asked to do many different kinds of activities with their child. The reading level is very high, too high for many of our families. Teachers did have to work hard with some children to establish patterns of effective return of homework. But most teachers who worked at it were able to achieve high rates of such return—rates that were surprising to them and considerably higher than they had experienced in the past.

Because many Latino families are family-centered, we designed games that family members could play to help their child practice important math competencies. Some families and some children had no previous experience in playing board games or games with cards, so teachers had family nights in which they taught families to play the games. These were generally well attended and enjoyed by all. One school librarian made fancier versions of the games and made them available to family members in the school library so that they could learn the games there. The feedback from families involved in using the games was quite positive. We also suggested activities parents could do to help children see math in their everyday lives. Some teachers also brought in family or community members to illustrate math links to the real world (Civil 1992, 1993; and Moll et al. 1992 also did this).

USING THE EQUITY PEDAGOGY

Although our teachers had the support of teaching materials we were developing in the classroom, teachers can implement many aspects of TMEP while using many different kinds of teaching materials. We have three major suggestions for teachers. First, try to collaborate with at least one other teacher in this effort. This collaboration can stimulate ideas and provide support for the different kinds of changes each teacher will be making.

Second, concentrate efforts on a few central grade-level goals. Try to help the classroom become a place where everyone explicitly is helping everyone learn those central goals. For these major goals, use all six aspects of TMEP (see Table 14.1).

Third, devise learning trajectories for these central goals by paying attention to children's different solution methods and to their errors. Everyone in the class can become a solution collector and an error detective who finds and helps correct errors. This approach removes the total responsibility from the teacher to identify, understand, and correct errors. It also provides more-advanced children important opportunities for rich mathematical learning as they compare and contrast different methods.

CONCLUSION AND A VISION OF THE FUTURE

Equity requires a balance between rights and responsibilities, between the needs of less-advanced and more-advanced children, and between the needs of less-advantaged and more-advantaged children. Equity guarantees access to mathematical learning, but requires and supports continuous effort by all participants. Meaning-making and problem solving are major foci in an equity classroom. These are facilitated by children's use of MDMs, situations linked to mathematics language and symbols, and rich language use. Practice focused on vital prerequisite topics also plays a crucial role. We offer our vision of a combination of the best elements of the twentieth century in the hope that it can stimulate productive dialogue and action that can move us toward achieving more substantial mathematics learning by all in the twenty-first century.

Such productive dialogue and action could focus on many elements of the equity pedagogy. The ease of communicating through e-mail and the Web, combined with technological advances that could make sharing videotaped segments of classroom interactions inexpensive and widely available, could enable teachers and mathematics educators all over the country, and indeed all over the world, to share productive conceptual supports. Ways to improve classroom mathematical discussions could be shared, and participants could share, critique, and suggest improvements for a given videotaped segment of classroom discussion. Children's language and solution methods could be shared, as could different traditional methods coming from homes, especially those of families from other countries. Errors children make could be shared, and other teachers could participate in debugging those that are not obvious to a given teacher. Clearly, any of these kinds of interactions could also take place in face-to-face teacher groups.

TMEP focuses on meaning-making, high expectations, and a collaborative culture for supporting understanding of children. Teacher development efforts for both in-service and preservice teachers require the same features. Curriculum developers, administrators, in-service educators, and teachers themselves working alone or in groups need to start where teachers are and keep learning meaningful. They must work to develop a collaborative math talk culture of understanding, explaining, and helping about teaching math with understanding. We have found that much teacher learning occurs in a classroom that uses TMEP through mutually stimulating cycles of teacher and child mathematical learning (Drake, Spillane, and Hufferd-Ackles in press; Hufferd-Ackles 1998; Hufferd-Ackles and Fuson 1999). Teaching using TMEP leads to teacher growth that is stimulating and fulfilling as well as to increased understanding by all children.

REFERENCES

Belenky, Mary F., Blythe M. Clinchy, Nancy R. Goldberger, and Jill M. Tarule. *Women's Ways of Knowing: The Development of Self, Voice, and Mind.* New York: Basic Books, 1986.

Civil, Marta. "Entering Students' Households: Bridging the Gap between Out-of-School and In-School Mathematics." In *Proceedings of the 44th International Meeting of the International Commission for the Study and Improvement of Mathematics Teaching,* edited by A. Weinzweig and A. Cirulis, pp. 90–109. Chicago: International Commission for the Study and Improvement of Mathematics Teaching, 1992.

———. "Household Visits and Teachers' Study Groups: Integrating Mathematics to a Sociocultural Approach to Instruction." In *Proceedings of the Fifteenth Annual Meeting of the North American Chapter of the International Group for the Psychology of Mathematics Education,* edited by Joanne R. Becker and Barbara J. Pence, Vol. 2 (ERIC document ED372917), pp. 49–55. San José, Calif.: San José State University, 1993.

Cooper, Robert G. "The Role of Mathematical Transformations and Practice in Mathematical Development." In *Epistemological Foundations of Mathematical Experience,* edited by Leslie P. Steffe. New York: Springer-Verlag, 1991.

De La Cruz, Yolanda. "Reversing the Trend: Latino Families in Real Partnerships with Schools." *Teaching Children Mathematics* 5 (January 1999): 296–300.

Drake, Corey, James P. Spillane, and Kimberley Hufferd-Ackles. "Storied Identities: Teacher Learning and Subject Matter Context." *Journal of Curriculum Studies.* In press.

Fuson, Karen C. *Latino Children's Construction of Arithmetic Understanding in Urban Classrooms That Support Thinking.* Paper under revision, 1999.

———. "Research on Learning and Teaching Addition and Subtraction of Whole Numbers." In *The Analysis of Arithmetic for Mathematics Teaching,* edited by Gaea Leinhardt, Ralph T. Putnam, and Rosemary A. Hattrup, pp. 53–187. Hillsdale, N.J.: Lawrence Erlbaum Associates, 1992a.

———. "Research on Whole Number Addition and Subtraction." In *Handbook of Research on Mathematics Teaching and Learning,* edited by Douglas Grouws, pp. 243–75. New York: Macmillan, 1992b.

Fuson, Karen C., Yolanda De La Cruz, Stephen B. Smith, Ana María Lo Cicero, Kristin Hudson, and Pilar Ron. *Toward a Mathematics Equity Pedagogy: Creating Ladders to Understanding and Skill for Children and Teachers.* Manuscript in final preparation, 1999.

Fuson, Karen C., Kristin Hudson, and Pilar Ron. "Phases of Classroom Mathematical Problem-Solving Activity: The PCMPA Framework for Supporting Algebraic Thinking in Primary School Classrooms." In *Employing Children's Natural Powers to Build Algebraic Reasoning in the Context of Elementary Mathematics,* edited by J. Kaput. Hillsdale, N.J.: Lawrence Erlbaum Associates, in press.

Fuson, Karen C., Ana María Lo Cicero, Pilar Ron, Yolanda De La Cruz, and L. Zecker. *El Mercado: A Fruitful Narrative for the Development of Mathematical Thinking in a Latino First- and Second-Grade Classroom.* Manuscript under revision, 1999.

Fuson, Karen C., and Stephen T. Smith. "Complexities in Learning Two-Digit Subtraction: A Case Study of Tutored Learning." *Mathematical Cognition* 1 (1995): 165–213.

———. "Supporting Multiple Two-Digit Conceptual Structures and Calculation Methods in the Classroom: Issues of Conceptual Supports, Instructional Design, and Language." In *The Role of Contexts and Models in the Development of Mathematical Strategies and Procedures,* edited by Meindert Beishviszen, Koeno P. E. Gravemeijer, and Ernest C. D. M. van Lieshout, pp. 163–98. Utrecht, Netherlands: Freudenthal Institute, 1997.

Fuson, Karen C., Stephen T. Smith, and Ana María Lo Cicero. "Supporting Latino First Graders' Ten-Structured Thinking in Urban Classrooms." *Journal for Research in Mathematics Education* 28 (1997): 738–60.

Hiebert, James, Thomas Carpenter, Elizabeth Fennema, Karen C. Fuson, Diana Wearne, H. Murray, A. Olivier, and P. Human. *Making Sense: Teaching and Learning Mathematics with Understanding.* Portsmouth, N.H.: Heinemann, 1997.

Hufferd-Ackles, Kimberly. "Student Contributions Correcting and Strengthening Teachers' Conceptions of Math." In *Proceedings of the Twentieth Annual Meeting of the North American Chapter of the International Group for the Psychology of Mathematics Education,* Vol. 2, edited by Sarah Berenson, Karen Dawkins, Maria Blanton, Wendy Coulombe, John Kolb, Karen Norwood, and Lee Stiff, pp. 541–48. Raleigh, N.C.: North Carolina State University, 1998.

Hufferd-Ackles, Kimberley, and Karen C. Fuson. "Students as Teachers: Strengthening Teachers' Understandings of Math." Paper presented at the Annual Meeting of the American Educational Research Association, Montreal, Canada, 1999.

Leinhardt, Gaea, Ralph T. Putnam, and Rosemary A. Hattrup, eds. *The Analysis of Arithmetic for Mathematics Teaching.* Hillsdale, N.J.: Lawrence Erlbaum Associates, 1992.

Lo Cicero, Ana María, Yolanda De La Cruz, and Karen C. Fuson. "Teaching and Learning Creatively: Using Children's Narratives." *Teaching Children Mathematics* 5 (May 1999): 544–47.

Lo Cicero, Ana María, Karen C. Fuson, and Martha Allexsaht-Snider. "Making a Difference in Latino Children's Math Learning: Listening to Children, Mathematizing Their Stories, and Supporting Parents to Help Children." In *Changing the Faces of Mathematics: Perspectives on Latinos,* edited by Luis Ortiz-Franco, Norma G. Hernandez, and Yolanda De La Cruz, pp. 59–70. Reston, Va.: National Council of Teachers of Mathematics, 1999.

Moll, Luis C. "Bilingual Classroom Studies and Community Analysis: Some Recent Trends." *Educational Researcher* 21 (1992): 20–24.

Moll, Luis C., C. Amanti, D. Neff, and Norma Gonzalez. "Funds of Knowledge for Teaching: Using a Qualitative Approach to Connect Homes and Classrooms." *Theory into Practice* 31 (1992): 132–41.

Murphy, Lauren. "Learning and Affective Issues among Higher- and Lower-Achieving Third-Grade Students in Math Reform Classrooms: Perspectives of Children, Parents, and Teachers." Ph. D. diss., Northwestern University, 1997.

Resnick, Lauren. "From Aptitude to Effort: A New Foundation for Our Schools." *Deadalus* 124 (1995): 55–62.

Ron, Pilar. "Spanish-English Language Issues in the Math Classroom." In *Changing the Faces of Mathematics: Perspectives on Latinos,* edited by Luis Ortiz-Franco, Norma G. Hernandez, and Yolanda De La Cruz, pp. 23–34. Reston, Va.: National Council of Teachers of Mathematics, 1999.

Secada, Walter G. "Race, Ethnicity, Social Class, Language, and Achievement in Mathematics." In *Handbook of Research on Teaching and Learning Mathematics,* edited by Douglas Grouws, pp. 623–60. New York: Macmillan, 1992.

Simon, Martin A. "Reconstructing Mathematics Pedagogy from a Constructivist Perspective." *Journal for Research in Mathematics Education* 26 (March 1995): 114–45.

Vygotsky, Lev S. *Thought and Language.* Edited and translated by Eugenia Hanfmann and Gertrude Vakar. Cambridge, Mass.: Massachusetts Institute of Technology Press, 1962. (Original work published 1934).

———. *Thought and Language.* Edited and translated by Alex Kozulin. Cambridge, Mass.: Massachusetts Institute of Technology Press, 1986. (Original work published 1934).

15
Exploring Mathematics through Talking and Writing

David J. Whitin

Phyllis Whitin

DEVELOPING a personal voice in mathematics is one of the most important goals for students of the twenty-first century. Two of the most powerful avenues for cultivating this personal signature are talking and writing (Countryman 1992; Schifter 1996; Whitin and Whitin in press). When these tools for expressing ideas are used in open-ended ways, they have the potential for inviting everyone into mathematical investigations. In classrooms of tomorrow all students must feel comfortable in joining the mathematical conversation. These goals are even more urgent now because of our increasingly pluralistic classrooms in which students bring a wide range of cultural and linguistic experiences. As teachers in the twenty-first century we must build mathematical communities that give everyone a place in our educational sun; we must broaden the eye of the needle so that it is possible for all children to be inquirers themselves and to exhibit a mathematical habit of mind (Eisner 1991).

The learning vignettes that we describe in this article are based on certain important principles of learning: children are constructors of their own mathematical knowledge (von Glasersfeld 1995), learning is a social process (Vygotsky 1978), and writing and talking are tools for reflecting thought as well as generating new thoughts (Murray 1968; Vygotsky 1978). The question then becomes, "How do we operationalize these principles of learning for tomorrow's children?" This article focuses on the significant role that teachers play in putting these beliefs into action.

Some of the specific teacher strategies discussed in this article include (1) supporting the art of problem posing so that children can gain confidence in formulating and testing hypotheses, (2) turning children's questions back to the group so that they develop more ownership for the problem-solving process, (3) keeping explorations open-ended so that students feel free to contribute a diversity of responses, (4) sharing personal mathematical connections so that children perceive teachers as partners in learning, and (5) playing the role of devil's advocate so that children can view their answers from a new perspective.

This article highlights on the potential of children's literature as a context for employing these strategies (Whitin and Wilde 1992, 1995). As a team of teacher-researchers we show how a single piece of literature can draw out the individual voices of students. We share the work of two fourth-grade classes that became involved in exploring such topics as prime and composite numbers, divisibility, and patterns.

Linking Language and Mathematics through Story

A Remainder of One (Pinczes 1994), a delightful division story told in verse, was the springboard for our activities in both fourth-grade classes. In the story, soldier ant Joe struggles to please the queen bee, who demands that a battalion of 25 ants march in equal rows. At first Joe tries to organize the battalion into 2 rows of 12, but finds that he is the outcast "remainder of one." Similarly, he is the lone remainder in the marching arrays of 3×8 and 4×6. Finally, the problem is solved when Joe and his comrades march in 5 rows of 5, and Joe finally takes his place in the triumphant march before the queen.

Ryan and Lauren used their knowledge of how stories are constructed to investigate some of the mathematical ideas in *A Remainder of One*. After hearing the story, Ryan commented that he thought ahead how the problem would be solved. When he realized that the author was adding a row each time, he predicted that the story would conclude with $5 \times 5 = 25$. Ryan's response showed us one of the benefits of reading mathematical stories to children. He knew that plots are usually problem-centered and that the resolution of the story would occur when the problem was solved. He combined his mathematical knowledge of multiplication with this understanding of story.

Lauren was intrigued by the author's purpose. She mused, "I wonder why the author picked 25?" As a writer and a reader, Lauren was aware that authors make choices in their writing in order to achieve certain purposes. She hypothesized that Pinczes had a specific purpose for choosing 25 for the number of ants. We teachers decided to turn Lauren's question back to the group for further consideration. In this way we not only honored Lauren's question but also gave the class the responsibility for exploring it together. An important role for teachers is to listen for intriguing questions and comments and use them to extend the mathematical investigation. This strategy is an important one for classrooms of the twenty-first century because it recognizes the right of students to set some of the directions of the investigation. Classrooms of the future must give children the opportunity to do what mathematicians do, and part of that opportunity includes raising their own questions.

We therefore asked these children, "Why did the author choose 25 ants for the story? What would have happened if she chose another number?" We phrased the question in this way in order to promote a sense of problem-posing. Teachers can help children explore the characteristics of a mathematical story by encouraging them to change the story's attributes in different ways. Part of appreciating the unique properties of the number 25 is to contrast it with those of other numbers. Alex suggested, "The author chose 25 so the book won't be over so soon." When asked to elaborate, he discussed the futility of using arrays composed of even numbers; the story would be over on the first page of two rows. Tony thought, "She probably picked 25 because it has 3 remainders, then 5, so she could make it not work, then work at the end." Carla thought about the same idea from a different perspective. Drawing on her recent work with prime and composite numbers, she suggested, "It needs to be a composite number." Her comment helped us ask the class another problem-posing challenge: "What would happen if the author used a prime number of ants?" The children realized that in this case, the ants could only march in one long column. In Carla's words, "Prime numbers would make the story boring, and no one would like the book." Here again it was the students' appreciation of the mathematics that helped them better understand the authoring decisions of the story.

Ryan's knowledge of how stories are put together helped him formulate his mathematical prediction, whereas Carla's knowledge of the mathematics helped her better appreciate the story structure. Thus, an important instructional strategy is to challenge children to alter the variables of a problem. It gives them ownership for problem formulation and allows them the opportunity to view the given problem in the context of other possible problems (Brown and Walter 1990).

This discussion piqued the children's interest in trying out some of the other numbers for themselves. Another important role for teachers is to keep the exploration open-ended to ensure a diversity of responses. We therefore gave the children time to construct different arrays with one-inch tiles. The children's writing about their explorations reflected their concern for the character Joe in the story. For example, Jeremy decided to try 12 tiles. Although he found that there was no remainder with 2, 3, or 4 rows, he was content. He wrote, "I notice that there is no way for Joe to be left out because twelve has lots of ways. Joe is lucky that he is not left out because in the book he was left out all the time except the end." From the point of view of the character Joe, an arrangement of 12 was a happier story! Selita, on the other hand, could find no solution for Joe with 43 tiles. When she tried 3 rows, she wrote, "I had 43 tiles and I arranged them like 3 tiles on the top and bottom and 14 tiles on the sides." She then drew the single tile as Joe outside the array ($3 \times 14 = 42$, R1). After trying 4 rows she again drew a single tile ($4 \times 13 = 42$, R1) but labeled it, "With a leader." She continued to write, "I wonder why I could not do 43—maybe because it is an odd number. The reason why I did it with a leader was because there was no way I could do 43. I could do 44 but not 43." Having the opportunity to use a different number of tiles, both Jeremy and Selita explored the issue of divisibility in their own way. Selita's observation about 43 being an odd number helped us see that she may not have sorted out the difference between odd numbers and prime numbers. Thus, the discussion that stories evoke can give teachers an effective way to assess children's mathematical understanding.

Whitney was also interested in composing a happy ending for Joe. Whitney loved to write. The initial appeal of composing her own story inspired her to investigate the mathematics of the situation. She chose a prime number, as Selita did, and wrote the following story for 13:

> There are two rows of 6 ants and there is one left over. "Move out of the way," says the QUEEN ANT. "You are a left over." The poor little ant went home and thought, what could I do to be in the march. Maybe I could make 3 rows of 4 ants. So the next day the poor ant went to ask his friends if they would make 3 rows of 4 ants, and they said, "Yes," but the poor little ant was still left over. So the poor little ant went home and thought. At last he thought of something, and he said, maybe I could try 4 rows of 3. So the next day the ant went to ask his friends if they would try 4 rows of 3 ants and they said, "Yes." But when they tried the poor ant was still left over. "Move," said the QUEEN BEE [sic], "You are still a left over." So the poor ant went home and thought. Maybe I could try a whole row. So the next day the ant went to ask his friends if they would try a whole row, and it worked, and the QUEEN BEE [sic] was very happy, so she gave the ant a BLUE RIBBON.

Whitney followed the same sequence of divisors as the original story and found three consecutive quotients with a remainder of one. However, she saw that the problem of the story was not going to be resolved by employing divisors greater than 4, and so she chose the only possible solution for a prime number, i.e., a single column. Her story helped show how the multiplication properties of 13 were similar to, but different from, 25.

These three children, like many of the students, were drawn to the human element of this story about soldier Joe. They were all concerned about Joe's plight and were eager to create stories that had a happy ending. Jeremy was delighted with 12 because its many factors made Joe a happy soldier all the time; Selita and Whitney demonstrated the problem with cumbersome prime numbers but they invented solutions to rescue Joe from this difficult situation of only two factors. Thus, the empathy expressed about Joe's predicament caused the children to find unique mathematical solutions to the problem. Their written stories show again the interesting link between story structure and mathematics. The instructional strategy of inviting these problem variations gave students the opportunity to explore further their knowledge of prime and composite numbers.

Linking Language and Mathematics through Talk

With another group of children we chose to focus on the uniqueness of 25 in a different way. We asked students to use some tiles and complete a chart (fig. 15.1) that showed various possible groups of ants and the results of other divisors. Teachers can help children look for patterns by suggesting charts and other tools for organizing information. Once the students had completed this chart, we held a discussion to share our initial observations. This discussion highlighted the potential of talk for exploring mathematical ideas. Teachers can capitalize on this potential by encouraging children to elaborate on their responses, justify their reasoning, and share their rough-draft thinking. Some important questions that teachers can ask include: "What did you notice? What patterns do you see? Why is that pattern occurring? Who else has another idea? Why would that make sense?" Asking open-ended questions such as these is an important instructional strategy. These questions broaden the range of possible responses. Teachers of the twenty-first century must structure conversations so that children of various mathematical abilities are invited to participate. This group of children had already had many mathematical discussions. The conversation that follows reflects this experience of exploring mathematical ideas in open-ended ways.

	Number of Rows			
Number of Ants	2	3	4	5
12	6	4	3	2R2
13	6R1	4R1	3R1	2R3
14	7	4R2	3R2	2R4
15	7R1	5	3R3	3
16	8	5R1	4	3R1
17	8R1	5R2	4R1	3R2
18	9	6	4R2	3R3
19	9R1	6R1	4R3	3R4
20	10	6R2	5	4
21	10R1	7	5R1	4R1
22	11	7R1	5R2	4R2
23	11R1	7R2	5R3	4R3
24	12	8	6	4R4
25	12R1	8R1	6R1	5

Fig. 15.1. Chart for dividing ants into rows

David, the university collaborator, began the discussion by asking, "What do you notice about the remainders on this chart?" One of the first observations focused on the uniqueness of 25. Several students saw that 25 had three quotients with a remainder of one (25 ÷ 2 = 12 R1, 25 ÷ 3 = 8 R1, 25 ÷ 4 = 6 R1). Playing the devil's advocate we teachers said, "But look at 13 and 21. They each have three quotients with a remainder of one (13 ÷ 2 = 6 R1, 13 ÷ 3 = 4 R1, 13 ÷ 4 = 3 R1; and 21 ÷ 2 = 10 R1, 21 ÷ 4 = 5 R1, 21 ÷ 5 = 4 R1). Why didn't the author pick one of those numbers?" This response challenged the children to look more closely at 13, 21, and 25 and describe more explicitly the unique attributes of 25, i.e., it had three consecutive remainders of one (21 did not) followed by a remainder of zero (13 did not). By playing devil's advocate, teachers can encourage children to avoid hasty conclusions, refrain from overgeneralizations, and look more closely to make finer distinctions. The chart proved to be a helpful tool in appreciating why the author chose 25 to tell this particular story.

Encouraging More Explanation and Description

This next part of the discussion highlights another crucial role that teachers play: challenging students to explain their thinking further. Joseph noted that under the divisor of 4 there were remainders of 1, 2, 3, and that then "it came out even." David responded, "And after a remainder of 3, what finally happens?" Joseph replied, "If there's not another number that couldn't be divided evenly and get a remainder of 3 or something, it will probably go to a full even number." Joseph's comments pointed to the cyclical pattern of remainders, a distinctive feature of modular arithmetic. Lori explained this pattern in another way: "Under the fifth row it only has remainder 1, 2, 3, 4; and under the fourth row it only had remainder 1, 2, 3, and under the third row it has remainder 1, 2, and under the first row it only has remainder 1." David replied, "Why do you suppose that is happening?" It is not enough for students merely to describe patterns; rather, an important role of the teacher is to challenge students to explain why that pattern is occurring. It is this followup question that often gets at the heart of the mathematics. Lori was not quite sure about the reason, but ventured, "Well, I'm just going to take a guess, but maybe since 5 is a odd number, maybe it comes out a certain way for the numbers." Since David wanted Lori to be more specific about her response, he focused her attention on one part of the chart. Sometimes it is important for teachers to narrow the focus of the problems and still allow children the freedom to draw their own conclusions. David said, "Look at the fifth row. There is a remainder of 1 for 16, a remainder of 2 for 17, a remainder of 3 for 18, a remainder of 4 for 19, but there is no reminder for 20. Why isn't there a remainder for 20?" Lori responded, "Because 5 times 4 is 20." In this case, David highlighted the divisor of 5 that Lori found interesting, but then focused on the specific numerical pattern found beneath that column. Then he asked Lori to explain it further. This kind of interchange between student and teacher marks the delicate tightrope that teachers must walk—preserving the autonomy of children's problem-solving efforts while at the same time providing enough guidance to highlight the mathematics of the situation.

Putting problems in front of the group is an important role of teachers. The implicit message is: if we can't figure things out on our own, we can figure things out with the help of others. In this way we generate solutions together. For instance, after Lori mentioned the issue of odd numbers, Stephanie contributed a new insight that focused on this same characteristic: "I notice that there are a lot more odd remainders than even remainders. I counted them, and there are 16 even remainders and 24 odd remainders." When Stephanie couldn't explain why this relationship was occurring, David invited other students to help. Alex explained the relationship in this way: "There aren't that many remainder 4s because it only happens on 5s. There's lots of remainder 1s because it get close to the number a lot." Jessica extended this fascination with odd and even numbers with another insight: usually there are more odd remainders than even remainders if the number (dividend) is odd, and there are more even remainders than odd remainders if the number is even. The children found 16 to be an anomaly: it was the only even number with all odd remainders. Thus, Stephanie's initial observation not only gave Alex the opportunity to explain the frequency of remainders but also provided Jessica with the opportunity to make an additional insight about the odd and even nature of the remainders. It is important for teachers to make children aware of this generative potential of talk. For instance, we can ask Jessica, "How did you get your idea for looking at odd and even numbers?" Often children will credit their peers. In this way children publicly acknowledge and validate the power of a collaborative community.

Another important instructional strategy is to encourage students to describe the same mathematical idea in different ways. This strategy often means allowing students to return to ideas that were previously mentioned in the discussion. This norm for conversing grants students the right to mull over past ideas and then offer alternative interpretations or explanations. For instance, the idea of modular arithmetic that Joseph raised initially was discussed later by several students. Jason explained, "In every number the remainder keeps getting one more.... You get more remainders with 5 because 5 is a bigger number. The bigger the number (divisor) gets, the more remainders you can have." Still later on in the conversation Watkins said, "I see a pattern under the 4 (divisor): 3, R1, R2, R3, 4, R1, R2, R3, 5. It's building up, and then it switches over." Several minutes later Jamie made this comparison: "Remainders are building up each time, getting higher and higher, getting closer and closer to building a family." Each student added a unique description of the same pattern: Jason offered a generalization, whereas Watkins and Jamie shared some visual images, such as remainders "building up" and "switching over" to become "families" (i.e., increasing the quotient by one more). By encouraging multiple explanations teachers not only validate the individual voices of children but also build a rich fabric of mathematical understanding. Stories have the potential for opening up these kinds of exploratory conversations; in this way all learners can contribute valuable interpretations to the group's collective understanding.

The Value of Sharing Rough-Draft Thinking

Encouraging students to share their rough-draft thinking is another useful strategy that builds productive classroom conversations. In this way the class can hear each other's most current thinking and thereby get a glimpse of people's thinking in process. For instance, Jamie said, "Every row down has four numbers that aren't remainders" (she probably first started to look under the divisor of 4). However, as she looked at the chart further she added immediately, "But it didn't work out." Nevertheless, Jamie's idea became the seed for another idea later in the discussion. Amanda shared, "I was going to add on to Jamie's [idea], but I think I have another idea. After every number there is a remainder." When asked to explain why this was happening, Lori pointed to the divisor of 3 and said, "You only get answers without remainders when you can count to that number by 3." Thus, it was Jamie's idea that focused on quotients without remainders that provided the catalyst for Amanda to look closely at the answers that followed those quotients. It is important for teachers to encourage students to salvage the discarded theories of others and resurrect those ideas in new ways. Children who are supported to be inquirers become skillful recyclers; they view abandoned ideas as fodder for building new relationships. As teachers, we can support children to take this inquiring stance by acknowledging the value of revising one's thinking. As teachers, we can respond to Jamie by saying, "Thank you for sharing that observation. You were wondering about a pattern, and then you looked more closely and discovered that it didn't work out. Sharing what's inside your head helps give all of us ideas." In this way children are valued as risk takers and creative problem solvers.

In a collaborative mathematical community the teacher is also an active participant in rough-draft thinking. When teachers join the conversation they demonstrate that they are learners themselves; this stance toward learning provides one of the most powerful lessons that we can give our students. In this way students see that all members of a collaborative community benefit from exploratory conversations. During this discussion about the remainder chart Kimberly shared an observation that sparked a new connection for David, the university collaborator. Kimberly contributed, "On 23, and on some other numbers, they all have remainders on them. So some numbers across you never get a number [without a remainder]. But the rest of the numbers you always get at least one number." Kimberly's comment highlighted the idea of prime and composite numbers, a topic that we had discussed earlier in the year and could now see again in another context. However, Kimberly's observation caused David to see a new relationship. He said, "Kimberly, your idea made me think of something. When you were talking about numbers and their remainders I noticed that there were two numbers that didn't have a remainder until the very end. They were 12 and 24. And then when I began to look at the numbers across for these two numbers I saw that each number for 12 doubled to make the corresponding number for 24. So I began to notice all of this because of what you said." It is important for teachers to make their thinking public so that they, too, are perceived as coexplorers in mathematics; it is equally important that they publicly recognize their intellectual indebtedness to their students. Classroom communities grow when everyone is teaching and everyone is learning. Talk is one way to support this collaborative learning.

Conclusion

The benefits of talking and writing are quite evident to the students who have experienced their potential. For instance, one day we said to the students, "In this class we often ask the question, 'Who has another idea?' Why do you suppose we keep asking that question?" Nia wrote this response in her journal: "Both of you not only want answers or ideas from that one person, but you want answers and ideas from every person, so you know what they are thinking, too. You want to know why we said that, and to know what is going on inside our brains. We can get a lot more ideas put together with 20 students rather than just asking one person." Students also recognize the potential of writing. Chris wrote, "When I write in math I understand things better;" and William noted, "You get more ideas. You get your imagination going."

Writing and talking cannot be scattered episodes in the school year, but must instead be part of the daily fabric of classroom activity. Children's literature is only one context in which this writing and talking can occur. The larger issue for teachers, as well as parents, means asking questions for which they do not already know the answers. "What do you notice?" and "What do you find interesting?" are two such questions that challenge adults to be more attentive listeners to children's thinking. For mathematics educators it means inviting prospective teachers to use writing and talking themselves to explore mathematical ideas. Teachers, too, must live the lives of mathematicians by posing hypotheses, searching for patterns, and altering problem variables. In these ways writing and talking can become powerful tools for building the kind of inclusive mathematical communities that we must have for the twenty-first century.

References

Brown, Stephen, and Marion Walter. *The Art of Problem Posing.* Hillsdale, N.J.: Lawrence Erlbaum Associates, 1990.

Countryman, Joan. *Writing to Learn Mathematics.* Portsmouth, N.H.: Heinemann, 1992.

Eisner, Elliott. "What Really Counts in Schools." *Educational Leadership* 2 (1991): 10–17.

Murray, Donald. *A Writer Teaches Writing: A Practical Method of Teaching Composition.* Boston: Houghton Mifflin, 1968.

Pinczes, Elinor. *A Remainder of One.* Boston: Houghton Mifflin, 1994.

Schifter, Deborah. *What's Happening in Math Class? Vol.1: Envisioning New Practices through Teacher Narratives.* New York: Teachers College Press, 1996.

von Glasersfeld, Ernst. *Radical Constructivism: A Way of Knowing and Learning.* London: Falmer, 1995.

Vygotsky, Lev. *Mind in Society.* Cambridge, Mass.: Harvard University Press, 1978.

Whitin, David, and Sandra Wilde. *Read Any Good Math Lately?* Portsmouth, N.H.: Heinemann, 1992.

———. *It's the Story That Counts.* Portsmouth, N.H.: Heinemann, 1995.

Whitin, Phyllis, and David Whitin. *Math Is Language Too: Talking and Writing in the Mathematics Classroom.* Urbana, Ill.: National Council of Teachers of English, 2000.

16

Unfinished Business
Challenges for Mathematics Educators in the Next Decades

Jeremy Kilpatrick

Edward A. Silver

As the twentieth century began, mathematics education in North America was just beginning to emerge as a field of serious study. Strong, forthright recommendations from the Committee of Ten on Secondary School Studies (National Education Association 1894) had spurred efforts on both sides of the United States–Canadian border to reshape elementary and secondary school mathematics programs. The College Entrance Examination Board, founded in 1900, and a committee appointed in 1902 by the fledgling American Mathematical Society were also attempting to make the secondary mathematics curriculum more uniform, encouraging schools to prune deadwood from, as well as to rethink, the mathematics to be taught. A protracted controversy, together with a flood of research into the mathematics that adults use, was triggered by Frank McMurry's 1904 address to the National Education Association in which he argued that the curriculum should be built on, and restricted to, such mathematics. Pioneering thinkers such as David Eugene Smith at Teachers College and J. W. A. Young at Chicago were attracting graduate students to the study of mathematics education, and dissertations on such topics as how children perceive number and space were beginning to appear. In 1908, the formation of the International Commission on the Teaching of Mathematics stimulated efforts in the United States and Canada to look at their school mathematics curricula and teacher-training practices. Despite these activities, however, there were no organizations, beyond some local clubs and several newly established regional associations, in which mathematics teachers could come together to examine and con-

---

We are grateful to George Stanic for his comments on a previous draft.

template their work. There were no North American journals devoted to mathematics education, and few books were being published that addressed issues and developments in the field.

By the end of the twentieth century, mathematics education in the United States and Canada had blossomed into a vast, intricate enterprise. The National Council of Teachers of Mathematics (NCTM), with about 110 000 members, was managing an extensive program of publications and meetings. Various organizations of a more specialized nature—for educators of mathematics teachers, Canadian researchers in mathematics education, mathematics teachers who have won national recognition, and the like—were meeting annually and publishing newsletters and journals for their members. Books and articles were pouring forth from innumerable outlets on all sorts of topics, from avoiding mathematics anxiety to learning calculus with the aid of graphing calculators. The field was marked not merely by these activities but also by university programs, degrees, departments, and faculty in mathematics education.

Mathematics educators, defined as anyone concerned professionally with the teaching and learning of mathematics at any level, accomplished much during the twentieth century. Students today benefit from a variety of instructional materials and activities engineered to help their learning. Teachers, for the most part, are better prepared mathematically and pedagogically than their counterparts in 1900 to engage their students with those materials and activities. Most school mathematics curricula are richer in topics and take these topics further than those of a century ago. Despite these accomplishments, however, one still hears many of the same complaints that people have been voicing for generations: Students aren't learning mathematics well enough; they leave school hating it. Teachers don't know enough mathematics and don't know how to teach it effectively. The school mathematics curriculum is superficial, boring, and repetitious. It fails to prepare students to use mathematics in their lives outside of school.

In this article, we consider some major challenges that mathematics educators face as the twenty-first century begins. These challenges are not unlike those dealt with in the past, but such challenges have persisted, mutated, and proliferated as both schooling and society have become more complex. If more students are to learn and use more mathematics more successfully than at present, the challenges we have identified must be met.

Ensuring Mathematics for All

Rapid growth in the North American school population marked the onset of the twentieth century, and the capacity of states and provinces to provide at least some secondary schooling for all students was severely tested. Throughout the nineteenth century, continued westward expansion and successive waves of immigration had fueled the demand for free public education across the continent. Many of the new entrants to the school population had cultural backgrounds and spoke languages different from those of their teachers, and these new students were generally perceived as far less capable than students from more familiar backgrounds. G. Stanley Hall (1904), a prominent psychologist of the time, saw a "great army of incapables" overrunning the schools. In response, many educators argued that school mathematics needed to be trimmed and tailored to suit the lesser capacities of the masses.

In the early decades of the century, educators tended to believe that children's intellect, although it could be exercised, set some fairly strong limits on what they could eventually learn. Thus, a popular idea of the time was that schooling could be made effective and efficient if students were sorted into the "capable" and the "incapable." In school mathematics, this meant that some students were judged likely to profit from instruction in mathematics beyond arithmetic; others were not. It was assumed that educators could unerringly detect the difference, perhaps with the assistance of objective, standardized tests.

It apparently did not occur to many people that the concept of *ability* might be questionable. As research over the last half century has shown, children said to lack ability may instead lack appropriate opportunities to learn or the support necessary to assist them in meeting learning expectations. Thus, the so-called *incapables* simply may not have encountered situations meaningful to them in which mathematics was important to know and in which they could turn to someone for help in understanding that mathematics. Proponents of ability-based sorting have likewise not been very concerned that the appraisal of ability might be difficult and subject to error. In fact, one's apparent mathematical ability may appear in some circumstances and not in others. Give a child a set of mathematical problems to be solved in ten minutes and graded for accuracy against the work of others, and the resulting performance may be dismal. Put the same child in a situation in which the problems are made meaningful, the same mathematics is used, and the solutions matter, and that child's performance can soar.

Today, educators are still challenged to find ways to provide mathematics for every student. And the notion that students can and should be sorted by mathematical ability continues to be widespread. As long as that ability is taken as a rock-solid property of the individual, however, it undermines a commitment to ensuring that all students receive an optimal education in mathematics. The inequities in mathematics achievement that are associated with the systematic sorting and tracking of students have been well documented (e.g., Oakes 1985). Thus, although the phrase "mathematics for all" has become a popular slogan among mathematics educators, and equitable mathematics learning is a central principle of *Principles and Standards for*

School Mathematics (NCTM 2000), significant challenges remain to be faced in the next century in order to make this slogan a reality. An educational system that has long rested many of its policies and practices on conceptions of innate mathematical ability will need to provide access for all students to high-quality mathematics in ways that will assist more of them to be successful in learning and using mathematics. A society that has tended to view mathematical ability as possessed by only a few select individuals will need to promote and support forms of instruction that help all to acquire high levels of quantitative literacy, skill in using mathematics, and appreciation of its nature and importance. And political leaders and ordinary citizens concerned about the quality of schooling will need to see that mathematics education nationwide can never be truly excellent unless there is excellent mathematics education in every classroom. Mathematics educators must play key roles in seeing that these challenges are addressed.

Promoting Students' Understanding

For mathematics educators, one of the most profound lessons of the past century rests on John Dewey's observation: We learn by doing and also by thinking about what we do. Many a student leaves school with a collection of well-practiced procedures and formulas but with only a hazy grasp of their meaning or of when they might be used. Students need more and better opportunities to understand the mathematics they are learning. They need good teaching. But what does that mean? Throughout the twentieth century, good mathematics teaching has been seen in various ways: giving learners clear explanations, identifying clear instructional objectives, prefacing instruction on complex knowledge and skills with hierarchical sequences of purported prerequisites, breaking instruction into small steps learners can easily take on their own, immersing learners in dilemmas with which they must struggle, helping learners resolve one another's confusions, tailoring instructional activities to individual learners' perceived ways of learning. Reconciling such disparate views requires that each be given a critical examination in the light of other modes of teaching.

An example is the tension between two models for planning and conducting lessons designed to promote understanding. In what might be called the *contingent* model, exemplified in vignettes in the *Professional Standards for Teaching Mathematics* (NCTM 1991), the path that teaching follows emerges during the lesson. The teacher's role is to orchestrate the discourse so that these students in this class will function as an intellectual community. The teacher sets up a situation and then responds to what the students are saying by building on their observations, seeking clarification, and challenging them to explain and justify. The goal is to help students develop their own and one another's understanding. In contrast, the *anticipant* model suggested by studies of teaching in some Asian countries follows a path carefully worked out in advance. Every lesson has a clear point to be reached. Because a given lesson has been tried out and refined many times by different teachers, a teacher can foresee these students' responses. The goal is to help students recognize, understand, and critique different ways of solving problems so as to improve their understanding. Both models make use of students' thinking: the contingent model as a rudder to steer the lesson toward an emergent goal; the anticipant model as a vehicle to reach a predetermined goal. The apparent conflict between these two approaches can be a starting point for mathematics educators' critical reflection. The challenge is not to ascertain which model is right and which is wrong or even which is better. Rather, it is to understand how each of these models works in helping students understand mathematics—especially the costs and benefits associated with each—and how the tension between the two might be resolved in any given instance.

Maintaining Balance in the Curriculum

Controversies over what mathematics students should learn, why they should learn it, and how it should be taught to them raged throughout the twentieth century. Goals such as training one's intellectual powers, preparing for the workplace, becoming an informed citizen, deriving esthetic satisfaction, helping one's country compete militarily or economically, and becoming confident in one's ability propelled arguments for bringing certain mathematical topics into the curriculum, keeping others there, and jettisoning still others. During the first quarter of the century, as noted above, arguments were advanced that the study of adult life would yield a clear picture of the mathematics that students needed to know and be able to do. Mathematics educators who prized their subject as one of the traditional liberal arts, however, resisted these arguments. Around mid-century, mathematics educators began to realize that no curriculum could be revised fast enough to keep abreast of social and technological change. No one could predict with any accuracy the mathematics that students would need when they became adults, even if one knew the careers into which they might be headed. Accordingly, curriculum developers proposed that school mathematics provide students with skills and understanding to help them learn as adults the specific mathematics they might need then. Abstract mathematical structures such as groups, rings, and fields appeared to provide the obvious foundation, and thus was born the "new math." By the end of the century, the argument was still being made that students needed to be prepared to learn mathematics as adults, but the curriculum had shifted to a much greater emphasis on applied mathematics as the best preparation. Adults were moving through a variety of jobs during their lifetimes, and the prediction problem had become even more intractable. Controversy continued as some mathematics educators began to question once more whether job preparation was the only, or even the best, reason for learning mathematics. Some wanted abstract mathematics restored to primacy in the school curriculum; others called for functional mathematics as the central goal.

In the twenty-first century, some of the main curriculum challenges concern balance. How are mathematics educators to balance the manifold goals that individuals and society have for school mathematics? How are the pure and applied sides of mathematics to be balanced? How is a balance between skill and understanding to be maintained? This last challenge was noted decades ago by William A. Brownell (1956) but has been made even more difficult to address with the advent of new technology that challenges the utility of paper-and-pencil skills.

The role that technology can and should play in skill development is an especially tough challenge, particularly since the long-term effects of the extensive use of computer and calculator technology are not known. Many mathematics educators and almost the entire general public assume that skill development must necessarily precede any use of such technology. That is clearly not the case, but the question of how to orchestrate technology use and skill development is far from being resolved, as is the question of how much skill it is useful to develop without using technology. Although opinions and anecdotes abound in regard to these issues, there has been too little deliberate reflection on hard evidence that might clarify and deepen the discussion.

Technology poses further challenges that go beyond skill development. Over the coming decades, students will make increasing use of technological tools in learning mathematics, using computers and calculators for a host of activities, including communicating, collecting and analyzing data, modeling real-world phenomena, manipulating mathematical expressions, and displaying information graphically. Without question, these tools can help students develop skills and understand mathematical ideas in new and different ways. But they are also fully capable of giving rise to curious misconceptions and, on occasion, profound misunderstanding. Swept along by a desire to update their instruction, mathematics educators have not given sufficient critical attention to the challenge of improving learning through appropriate uses of any technology, and not simply computer technology. Too often, technology is embraced as an unquestioned boon. Its limitations and disadvantages for mathematics instruction, as well as its potential for transforming the curriculum, have yet to be seriously questioned and analyzed.

A challenging aspect of achieving greater balance in the curriculum concerns the special qualities of mathematics compared to those of other disciplines. For example, attempts to achieve greater connections between mathematics and other school subjects have led mathematics educators to give greater attention to the inductive side of mathematics, to the ways in which induction is used in arriving at mathematical generalizations in much the same fashion as in biology or history. Also, efforts to develop students' skills in argumentation and in communicating their observations in mathematics have drawn on ideas from language and science. Teachers are now challenged to help students appreciate both the ways in which mathematics relates to their other work in and outside of school and the specific aspects of mathematics that distinguish it as a discipline, such as deductive proof and formalized abstraction and generalization. As students develop facility in conjecturing and testing their conjectures, how are they to be led to see that they do not yet have a proof? When mathematics is being used to answer practical questions, how can students come to know what a proof entails and why it might be needed? To ensure that students develop a sensible perspective on what mathematics is and what it can do, mathematics educators will need to give much more thought and effort to reconciling conflicting tensions in the curriculum.

Making Assessment an Opportunity for Learning

In 1993, the National Research Council proposed, among several principles for assessment, the Learning Principle: "Assessment should enhance mathematics learning and support good instructional practice" (p. 33). Two years later, NCTM (1995) echoed that principle with its Learning Standard: "Assessment should enhance mathematics learning" (p. 13). The argument in both documents was that assessment should provide not a time-out from learning but rather an opportunity for learning—for the teacher *and* for the student. Traditional tests and quizzes show the teacher, accurately or not, how students are doing, but it is the rare assessment in which students not only can reflect on their own understanding but also learn mathematics. The more artificial the assessment and removed it is from instruction, the less it can enhance learning. The more it resembles the situations in which mathematics is being learned and used, the more useful it becomes for everyone.

Changing assessment so that it can enhance rather than inhibit students' learning, however, poses a formidable challenge. Some aspects of the challenge lie in changing classroom assessment practices so that they not only shape instruction but also provide useful information for external assessors. A serious effort needs to be made to improve assessments mandated by authorities outside the classroom. Mathematics educators need to devote more attention to analyzing and critiquing such assessments so that they will neither intrude on nor conflict with students' learning. As long as mathematics educators simply advocate new forms of assessment but do not study and work to improve the conditions under which assessment is done, including its construction and subsequent uses, mathematics learning will be hindered.

Other aspects of the challenge of transforming assessment involve the students themselves. Reviewing the evolution of educational testing and assessment, Glaser and Silver (1994) argued that desired changes in assessment practices offer new opportunities for students' learning (p. 413):

> Closer ties between assessment and instruction imply that the nature of the performances to be assessed and the criteria for judging those performances will become more apparent to students and teachers. ... As performance criteria become more openly available, students will become better able to judge their own performance without necessary reference to the judgments of others. Instructional and assessment situations will provide coaching and practice in ways that help students reflect on their performances. Occasions for self-assessment will enable students to ... judge their own achievement and develop self-direction toward higher achievement goals.

When assessment is aligned with and integrated into instruction, it becomes a fertile opportunity for teachers to learn about what their students understand and what they can do. Because teachers are put off by assessment procedures that appear to demand too much time and expertise, another aspect of the assessment challenge is to convince teachers that when assessment is integrated into instruction, the time it takes is well spent. Yet another aspect is to develop the teacher's confidence. Too many teachers see themselves as needing to be given assessment instruments and told what and how to assess. Their ability to assess has been both undervalued and underdeveloped. In the next century, as was certainly true in the past, no one will be in a better position than a student's mathematics teacher to make sound judgments regarding the nature and extent of that student's mathematical accomplishments.

DEVELOPING PROFESSIONAL PRACTICE

Improving teachers' confidence and competence in their assessment activities is closely related to another crucial challenge faced by the community of mathematics educators as well as the larger society: changing the conditions under which teachers practice their profession. Most mathematics teachers work in relative isolation, with little support for innovation and few incentives to improve their practice. The possibility of collaborating with other teachers in developing instructional materials and assessment tools is typically absent. Many realize that they need to keep abreast of the field and to improve their preparation for teaching mathematics, but nothing in their workplace provides them with the requisite opportunities and resources.

Part of the challenge concerns the structuring of professional development systems to be more effective in enabling teachers to keep abreast of current developments and in touch with like-minded colleagues. Recent research suggests that teachers become better equipped to meet the kinds of challenges discussed above if they have opportunities to work together to improve their practice, time to engage in personal reflection, and strong support from colleagues and other qualified professionals (e.g., Smith in press). At present, most teachers' working conditions militate against reflective practice and thereby manage to defeat or attenuate repeated efforts at reform. Progress must be made in understanding how to design ways in which the practice of teachers—identifying mathematical goals, planning and conducting lessons, designing assessments, attending to students' thinking and learning, reflecting on goals and outcomes—can provide rich and powerful sites for teachers' learning (Brown and Smith 1997).

Serious attention to the conditions under which teachers do their work will help to create settings in which all students have a better chance to learn mathematics well. But there is another challenge that must be met. The mathematics teaching profession also needs to examine other ways to upgrade the professional knowledge and competence of its members. Too many students are enrolled in mathematics courses taught by individuals who lack what should be considered minimal preparation to teach the subject—namely, a major or a minor in mathematics at the undergraduate or graduate level. National data indicate that nearly one of every five secondary school students in the United States has a mathematics teacher who lacks this minimal qualification (National Center for Education Statistics [NCES] 1999). Moreover, there is an important distribution problem as well. Schools in poor urban or rural communities and schools serving high percentages of minority students are much more likely to have teachers who lack adequate subject matter preparation than are schools in affluent communities or those enrolling few minority students (NCES 1999). And the problem of teachers' preparation is especially severe in the elementary and middle grades, where many teachers have minimal content knowledge and lack both confidence and competence with respect to the subject. There are many reasons for the current problem, and the mathematics teaching profession shares the responsibility for its solution with many other parties, including school administrators, legislators and policymakers, and teacher educators. Yet there can be little doubt that the other challenges facing mathematics educators in the coming decades cannot be effectively addressed unless the challenges of developing their professional practice are also met.

THE IMPORTANCE OF REFLECTION

The title of this article, "Unfinished Business," is not meant to imply that the challenges we have identified will be completely and successfully met over the coming decades. On the contrary, we recognize that mathematics educators will always face the task of improving mathematics learning. Changing the learning and teaching of mathematics is not a technical problem; it involves, instead, a form of social change. It requires change not only in what students and teachers do but also in how they view their efforts and the circumstances under which they work. Clarion calls exhorting teachers to better practice and elegant printed materials showing the way will all be fruitless if the people involved in school mathematics—from students to teachers to administrators to parents and other caregivers to politicians—see no reason to change. But simply wanting to change in a certain direction is not enough. Social change requires that people support one another as they move toward a common, clearly understood goal. Coping with the challenges sketched in this chapter will require a critical stance in which the profession analyzes deeply, critiques thoroughly, and discusses vigorously those challenges. In a word, it requires *reflection*.

Reflection is an underused process for addressing the complexities of teaching and learning mathematics. Recent theories of how learners learn and how teachers teach have highlighted reflection as a powerful mechanism for refining thinking. People improve their thought and action by making their own mental processes the object of their thought and by changing those processes for the better. Learning has never been simply a matter of acquiring and retaining information; for information to become useful knowledge, it must be transformed by making it one's own, looking at it from all sides, seeing its interrelations, and thinking about its meaning. In other words, the learner must somehow reflect on his or her learning if it is to be put to use.

If the challenges identified above are to be tackled in a substantive way, reflection will need to be taken much more seriously and seen as applying not merely to individuals but to the entire field of mathematics education. For mathematics educators to form a true professional community, they need to engage in reflective practice. The developing profession of mathematics education in North America, as noted above, spawned numerous organizations, meetings, books, journals, curriculum development projects, curriculum and assessment materials, and professional activities, as well as many traditions of practice and many informal avenues of communication and dialogue. What the profession has not yet developed is a tradition of critical reflection on its own work. In such a tradition, everything mathematics educators do would be subject to careful appraisal. Practices both conventional and innovative would be open to question and discussion. Mathematics educators would consider critically the unexamined, and usually unexpressed, assumptions that guide much of their work. All types of research—action research by teachers, evaluations of innovative materials and practices, studies of basic teaching and learning processes—would become both a central activity of mathematics educators and a resource for improvement. Research would guide professional reflection and vice versa. In sum, the reflection that teachers encourage in students and that teacher educators encourage in teachers would become customary for the profession as a whole (Kilpatrick 1985, pp. 19–20). If this were the case, then authoritative documents, even those produced through consensus like NCTM's *Principles and Standards for School Mathematics,* would be taken as problematic rather than dogmatic.

PRINCIPLES AND STANDARDS: SACRED TEXT OR TOOLS FOR REFLECTION?

This view of reflection brings us to a final challenge for the next decade or so, this one for NCTM as a professional organization: promoting the use of its principles and standards not only as proposed solutions but also as tools for better understanding the nature of problems and challenges. For very good reasons, teachers look for reliable guidance wherever they can. Anyone concerned about helping students understand, value, and use mathematics will eagerly seek assistance from various sources. Professional organizations such as NCTM are happy to respond. But the conditions of learning and teaching in a specific classroom make it impossible for anyone outside that classroom to provide specific advice. Advisory statements about learning or teaching have to be understood as necessarily indeterminate and in need of interpretation.

In its "Standards 2000" document, NCTM (2000) puts forth a vision of what school mathematics might be in the coming decades. Many people will look to the document as either a sacred text or a guidebook that answers complex questions about what to teach, how to teach, and how to assess so that learning will be enhanced and students will be comfortable and confident with the mathematical power they acquire. That way of looking at *Principles and Standards for School Mathematics,* however, is less likely to be helpful to mathematics educators than is a view of the document as a tool for use in developing a reflective practice. The document does provide guidance, but it also identifies directly or indirectly many issues that mathematics educators face, thereby illuminating crucial sites for professional work. In so doing, it can serve as a springboard for professional reflection, discussion, and debate.

The vision outlined by NCTM needs neither endorsement in the form of hucksterism and mindless cheerleading nor critique in the form of mudslinging and unprincipled harangue. Mathematics educators need to engage in constructive dialogues about the vision of teaching, curriculum, and assessment that NCTM has offered. They need to resist clinging to a single view, becoming overly defensive, and disdaining further change without critically examining its costs and benefits. Any vision of school mathematics teaching and learning needs to be subjected to informed criticism. Moreover, it needs to change continually in light of the professions' experience and the better understanding it can achieve through a fair, thorough, and tough-minded debate.

If mathematics educators can adopt a more critical stance toward their work, there is good reason to be optimistic that many of the challenges of the next few decades can be met in ways that will lead to more effective professional practice. As noted above, the "business" will always be unfinished, but a strong commitment to steady, incremental change through a process that involves both action and reflection on action can ensure that continual progress is made toward improved mathematics learning by all students.

REFERENCES

Brown, Catherine A., and Margaret S. Smith. "Supporting the Development of Mathematical Pedagogy." *Mathematics Teacher* 90 (February 1997): 138–43.

Brownell, William A. "Meaning and Skill: Maintaining the Balance." *Arithmetic Teacher* 3 (October 1956): 129–36.

Glaser, Robert, and Edward A. Silver. "Assessment, Testing, and Instruction: Retrospect and Prospect." In *Review of Research in Education, Vol. 20*, edited by Linda Darling-Hammond, pp. 393–419. Washington, D.C.: American Educational Research Association, 1994.

Hall, G. Stanley. *Adolescence: Volume II.* New York: D. Appleton, 1904.

Kilpatrick, Jeremy. "Reflection and Recursion." *Educational Studies in Mathematics* 16 (1985): 1–26.

McMurry, Frank. "What Omissions Are Desirable in the Present Course of Study, and What Should be the Basis for the Same?" In *Journal of Proceedings and Addresses of the Forty-Third Annual Meeting.* Winona, Minn.: National Education Association, 1904.

National Center for Education Statistics, U.S. Department of Education. *Teacher Quality: A Report on the Preparation and Qualifications of Public School Teachers.* Washington, D.C.: U.S. Government Printing Office, 1999.

National Council of Teachers of Mathematics. *Assessment Standards for School Mathematics.* Reston, Va.: National Council of Teachers of Mathematics, 1995.

———. *Principles and Standards for School Mathematics.* Reston, Va.: National Council of Teachers of Mathematics, 2000.

———. *Professional Standards for Teaching Mathematics.* Reston, Va.: National Council of Teachers of Mathematics, 1991.

National Education Association. *Report of the Committee of Ten on Secondary School Studies, with the Reports of the Conferences Arranged by the Committees.* New York: American Book, 1894.

National Research Council. *Measuring What Counts: A Conceptual Guide for Mathematics Assessment.* Washington, D.C.: National Academy Press, 1993.

Oakes, Jeannie. *Keeping Track: How Schools Structure Inequality.* New Haven, Conn.: Yale University Press, 1985.

Smith, Margaret S. "'Balancing on a Sharp, Thin Edge': A Study of Teacher Learning in the Context of Mathematics Instructional Reform." *Elementary School Journal,* in press.

Index

American Mathematical Society (AMS), 132, 223
American Mathematical Association of Two-Year Colleges (AMATYC), 97
AMAYTC Standards, 133–34, 152–53
Appreciation of mathematics, 25
  beauty of mathematics, 6, 10, 134–35, 227
  enjoy mathematics, 6, 10
  hate mathematics, 10
  nature of mathematics, 17
  part of culture, 18, 29
  role in society, 17
  utility of mathematics, 21, 134, 135
Assessment
  external, 229–30
  importance of, 201
  learning from, 229–30
  of mathematics curriculum, 104–5
  of Web resources, 76
  professional development needs, 160, 230
  standards, 100

Back-to-basics movement, 5, 38
Brownell, William, 2, 228

Cambridge Conference on School Mathematics
  Goals for School Mathematics, 99
College Entrance Examination Board (CEEB)
  founding of, 223
  Program for College Preparatory Mathematics, 99
Conference Board of the Mathematical Sciences (CBMS), 38, 42
Connections (interdisciplinary), 17, 26, 29, 30, 32, 34, 36

Dewey, John, 132, 226

Equity in mathematics education, 3–4, 7, 13, 131, 141, 197–98, 200, 208–9
  inequities of teacher shortages, 231
  Mathematics Equity Pedagogy Program, 198 (*See also* Mathematics for all students)
Functional mathematics, 127–48, 228
  blend of traditional and reform, 135
  list of topics, 153–57

Hall, G. Stanley, 225
Hutchins, Robert Maynard, 132

International Commission on the Teaching of Mathematics, 1, 223

Language of mathematics, 134, 136
  communication, 6, 84
  construction of , 201, 204
  in bilingual classrooms, 198, 200, 206–7
  notations and representations, 47, 204–5
Learning mathematics, 133
  as sense-making activity, 115–16, 209
  barriers to, 4, 198
  constructivism, 147, 213
  context-based, 118–19
  for mastery, 147
  for systems thinking, 136, 145
  for transfer, 2
  for understanding, 2, 116, 141
  purpose of , 2, 16, 17, 18, 29, 98, 133, 134, 227
  stimulus-response theory, 2
Literacy
  mathematical, 3, 120
  quantitative (or numeracy), 24–25, 28, 32, 35, 129
  statistical, 186

Mathematical applications, 16, 23, 143–47
  as making connections, 119–20
  embedded in objects around us, 23
  technology facilitates, 23
  to other disciplines, 29–34
Mathematical Association of America (MAA)
  National Committee on Mathematical Requirements, 97
Mathematical (discourse) communities, 94, 213, 220
  cross-cultural, 82, 87
  classroom as, 209
  extranets, 83–94
  methods for developing, 188, 189–95, 199, 213–17
  supporting equity, 208–9
  Web-based, 83, 189, 195
Mathematical Sciences Education Board (MSEB), 107
Mathematics
  algebra, 1, 3, 18, 47, 60, 96, 98, 100, 103–4, 107, 108, 121, 137, 144, 147, 153, 156
  computation, 16, 154
  calculus, 2, 5, 10, 19, 20, 24, 52, 54, 57, 60, 74, 96, 103, 121, 128, 130, 134
  geometry, 18, 19, 21, 56, 69, 98, 103, 108, 154
  measurement, 154
  modeling, 84, 113–15, 156, 201
  problem solving, 5, 9, 16, 43–45, 55, 68, 119, 133, 139
  reasoning and proof, 5, 16, 31, 32, 42, 43–45, 85, 104, 105, 116, 154, 156, 229
Mathematics curricula and innovative programs, 6, 26, 29, 113, 116, 119, 123, 125–26
Mathematics education
  as lifelong activity, 10, 13
  challenges for, 224–34
  conditions for change in, 232
  history of, 1–3, 10, 223–24
  improving, 13–14
  teacher shortages, 231
Mathematics for all students, 3, 6, 9, 23, 28, 29, 34, 98, 120, 124, 133, 224–26
  some mathematics not for all, 21, 23, 25, 31
  tracking, 225
Mathematics for the workplace, 3, 16, 101–2, 120, 127–48, 227
  vocational mathematics, 129
Math wars, 128
McMurry, Frank, 97, 107, 223
Moore, Eliakim Hastings, 96–97, 107, 132

National Advisory Committee on Mathematics Education (NACOME), 38, 99
  Overview and Analysis of School Mathematics: Grades K–12, 100
National Assessment of Educational Progress (NAEP), 12, 51, 57, 103, 104–5, 130
National Commission on Excellence in Education (NCEE)
  A Nation at Risk, 38, 131
National Council of Supervisors of Mathematics (NCSM), 5
National Council of Teachers of Mathematics (NCTM), 5, 6–7, 31, 37–49, 84, 100, 107, 112, 128, 224
  Agenda for Action, 5, 38
  Assessment Standards for School Mathematics, 100
  Commission on Post-War Plans, 3, 98
  Commission on Standards for School Mathematics, 38
  Commission on the Future of the Standards, 41–42
  Professional Standards for Teaching Mathematics, 38, 100, 188, 226
  Standards 2000 Project, 41
  Standards 2000 Writing Group, 41–49
NCTM *Curriculum and Evaluation Standards for School Mathematics,* 6–7, 31, 37–49, 57, 97, 100, 112, 133, 151–52
  as framework for systemic reform, 39
  as policy tool, 39, 44–45
  as tool for teachers, 44–46
  as work-in-progress, 7, 39–40
  Content Standards, 43–44

criticism of, 40, 128
history of, 37–49
Process Standards, 40, 43–44, 133
research on impact, 39
NCTM *Principles and Standards for School Mathematics*, 7, 37–49, 84, 112, 158, 225–26, 233
  Association Review Groups (ARGs), 41–42, 48
  as tool for teacher discussion and reflection, 44–46, 60, 233
  Discussion Draft, 42–49, 100, 106–7
  research behind, 42
National Defense Education Act, 4
National Education Association, 97
  Committee of Ten on Secondary School Studies, 2, 223
National Institute of Education (NIE)
  Basic Mathematics Skills and Learning Conference (Euclid Conference), 5, 100
National Research Council (NRC), 48, 100, 107, 229
  Everybody Counts: A Report to the Nation on the Future of Mathematics Education, 100
National Science Foundation (NSF), 4, 39, 76, 79, 99, 107, 113
  history of, 98–99
  NSF-sponsored curriculum development projects, 39, 100, 113–25
  Systemic Initiatives, 39
Nature of mathematics
  as a collection of rules, 19
  as a construction of the human mind, 19
  as a creative medium, 21–23
  as a means for making visible the invisible, 19–21
  as a science of patterns, 19, 140
  as a way of knowing or explaining, 16, 18, 115
New Math, 4, 227
  criticism of, 5

Professional development
  as reflective practice, 231–33
  effective, 53–54
  implications of technology for, 60
  need for collaborative practice, 230
  research on, 231
  Teachers Teaching with Technology Program, 53–54
Progressive Education Association, 3
Progressive education movement, 3

School mathematics curriculum
  a mile wide and inch deep, 46, 136
  compartmentalization, 2, 96–97
  context-based, 68–73, 102, 137, 140
  course-taking patterns, 103
  focus on utility, 3, 102
  growth in twentieth century, 1–2, 100
  integration of topics, 2, 5, 13, 97, 108, 119–20
  national curriculum, 6
  need for balance in, 26, 59–60, 129, 132, 137, 209, 228
  needs of twenty-first century, 13, 108, 134, 153–57, 171–73
  ordering of topics, 124, 137
  organized around themes, 137
  traditional secondary framework, 100, 102–4, 129
School Mathematics Study Group (SMSG), 4, 99
Science and Public Policy Report (Steelman Report), 98–99
Second International Mathematics Study (SIMS), 100, 103
Skills
  basic, 6, 7, 8, 11, 46, 102, 118, 147
  early research on , 3
  knowledge, 203
  needed for future, 9
  occupational, 139
  paper-and-pencil, 54–60, 228
Smith, David Eugene, 1, 223
Sputnik, 4
Standards
  AMATYC Standards (*See* AMATYC)
  in response to societal demands, 31
  in relation to individual needs, 31–34
  NCTM Standards (*See* NCTM)
  necessarily indeterminate, 40, 233

state frameworks and textbook guidelines, 7, 39, 41, 105–6, 133, 152
universal standards and ecological parallel, 28, 31, 132
Statistics
  association between variables, 163–67
  data analysis, 85, 88–93, 136, 155, 159–62, 178–86
  experiments, 170–77
  inference, 156, 168–70
  resampling (bootstrap methods), 168–70
  simulations, 77, 168–70
  surveys, 170–71

Teacher preparation, 4, 14, 25, 35, 79, 98, 120–22
Teaching methods
  art of questioning and problem posing, 88–89, 213, 217
  calculator impact on, 56
  for appropriate use of technology (*See* Technology)
  for building discourse
    use of children's literature and storytelling, 201, 214
    Explain, Build On, Go Beyond strategies, 188–94
  for exploratory data analysis, 174–86
  for functional mathematics, 138–42
  for reaching disadvantaged, 198–200
  for working with families, 207–8
  models for teaching
    anticipant model, 226–27
    contingent model, 226–27
  teacher as guide or facilitator, 3, 97
  using linked multiple representations, 56
Technology, 7, 11, 55–56
  calculators/computers, 8, 17, 51–64
    appropriate use of, 9, 11–12, 58–61, 123, 175–78, 228

  history of, 52
  research on use of, 12, 57–58
  impact on
    access issues, 53
    basic skills, 57–58, 129–30
    learning priorities, 140
    mathematics curriculum, 54–55, 55–56, 97
    student needs, 102
    teaching methods, 60
    textbooks, 104
  supports
    lab approaches to mathematics, 140
    problem solving, 58
    understanding, 141, 148
    visualization, 145
  tools and software, 56, 59, 62–63, 71–75, 73–75, 122–23, 144, 175–78
  the Web, 67–80
    digital libraries, 68, 76, 78–79
    Internet, 63
    resource discovery, assessment, and reliability, 75–77
Third International Mathematics and Science Study (TIMMS), 46, 58, 64, 100, 103, 120, 131
  TIMSS Video Study, 105
Thorndike, E. L., 2, 3

U.S. Department of Labor
  What Work Requires of Schools (SCANS Report), 134, 153
University of Illinois Committee on School Mathematics (UICSM), 4, 99 (*See also* New Math)
Vygotsky, Lev , 201

Whitehead, Alfred North, 8

Young, J. W. A., 223